THE ENCYCLOPEDIA OF Sportscars

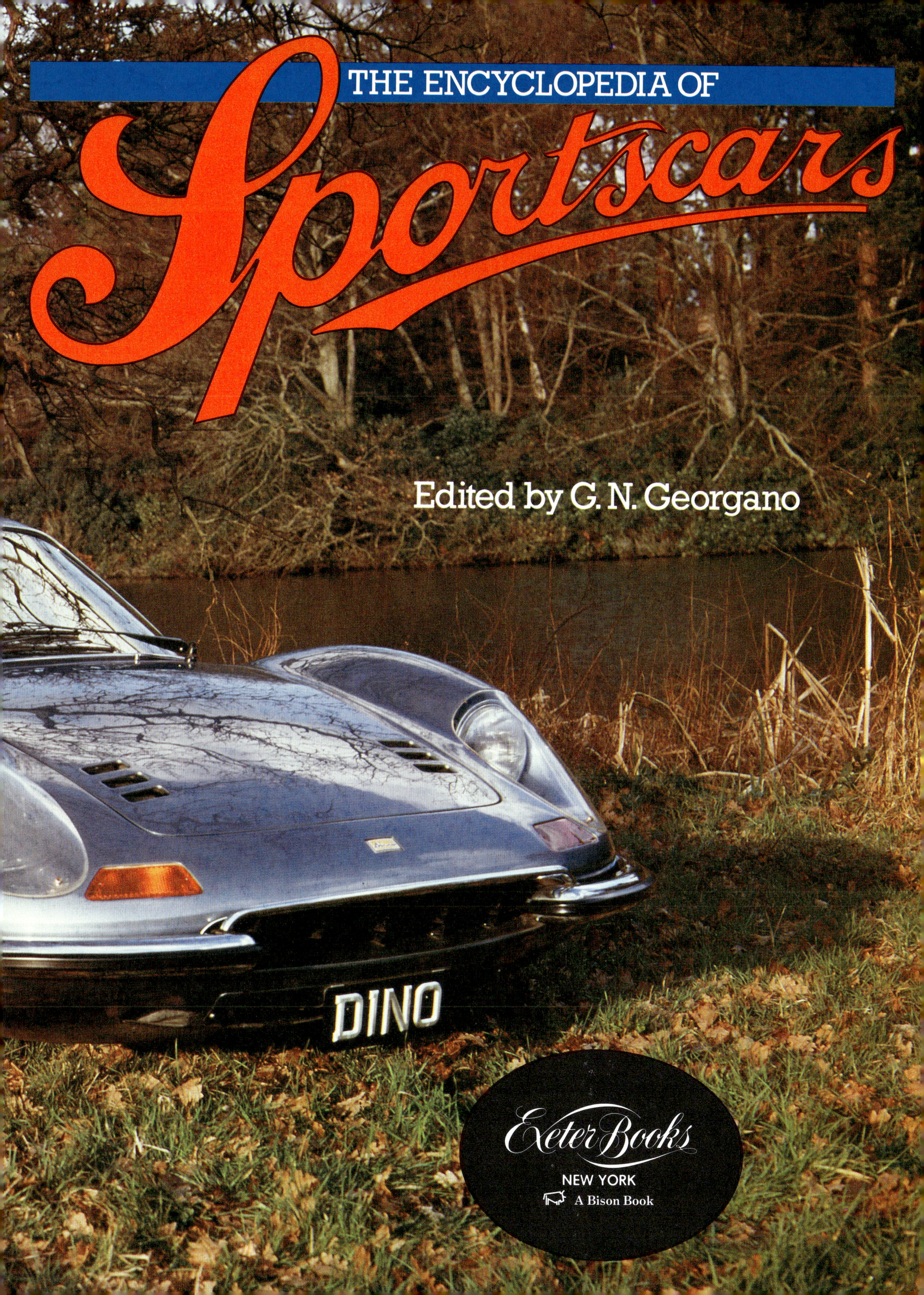

THE ENCYCLOPEDIA OF

Sportscars

Edited by G. N. Georgano

Exeter Books
NEW YORK
A Bison Book

First published in USA in 1985 by Exeter Books.
Distributed by Bookthrift. Exeter is a
trademark of Simon & Schuster, Inc. Bookthrift
is a registered trademark of Simon & Schuster,
Inc. New York, New York.

Produced by Bison Books Corp. PO Box 7608
Greenwich CT 06830

ISBN 0-671-07579-9

Printed in Belgium

Contents

Introduction

Many words have already been written in trying to define the term 'sportscar,' but still there is controversy, and disagreement. In this introduction, we will try to give a brief account of the evolution of various types that the term encompasses, and leave you, the reader, to agree or disagree with our choice.

For almost as long as there have been motorcars, people have been trying to make them better, and more pleasurable, to drive. Once the basic controls were mastered, the more go-ahead drivers then wanted to improve their speed and controllability.

In the 1890s, it was enough to spend a great deal of money, just to be able to drive from one place to the next, as quickly as one might ride a horse. Then came the developments of four-cylinder engines, Ackermann steering, by wheel instead of tiller, and pneumatic tires. As a result of this, and other change, speeds increased dramatically in the early 1900s, and motor racing became established, as a vital ingredient in the quest for more speed.

Cars were expensive luxuries, for the very rich, and it was not surprising that these owners mostly wanted to experience the maximum that their new purchases could offer. The older generation tended to retain their horses, but their wealthy sons adopted motoring as the latest, and most exciting, craze.

Soon a group of rich young men, their semi-professional employees and mechanics, began dicing for national, and then international, trophies. The cars they used – Mercedes, Mors, and Napier among them – made big concessions to the new sport, with the lack of wind-dragging weather protection, and by having the largest possible engines in flimsy chassis.

These expensive racing behemoths soon began to be supported by the factories, whose bosses saw success as their best chance of promoting normal road-going cars. Once the factories became seriously involved, however, the average motorist saw that he stood less and less chance of success against the 'works' drivers, so began to look for more amateur sport.

By 1905, in any case, racing cars were already moving away from the mainstream of automotive design, and becoming less and less practical for road use. The fast touring cars which evolved, that were eventually to develop into sportscars, bridged the gap between the racers, and the smooth and more sophisticated motor carriages.

Their practicality was proved on events like the German Herkomer, and Prince Henry Tours, held between 1906 and 1911. When the last Herkomer (of 1907) was held, one of the participants was Prince Henry of Prussia (the Kaiser's younger brother), and it was he who then organized the famous event which bore his name. This did much to engender friendly sporting competition, and to prove the worth of the new breed of strong and powerful cars, which had reasonable weather protection, and dust-diminishing bodywork.

Those who wanted cheap thrills had to turn to motorcycles, and it was from these machines that the first truly basic sportscars came around 1910. These were the most powerful types of cycle car, that could be bought – or built – for a pittance. Having motorcycle engines in dimunitive chassis, they gave all the performance and excitement, on a par with the two wheels.

There were dozens of contenders, but only a few of them became well known; one of the very best was the Morgan three-wheeler. Sporting cycle cars were particularly popular in Britain, and in France, where Marcel Violet started the first of numerous bizarre creations, many floors above the streets of Paris. When his first car was complete, he then had to dismantle it to get it down to the road!

America was less enthusiastic at first, for there were already plenty of cheap cars, like the Model T Ford, that were capable of reasonable performance and reliability. Before World War I, in fact, the Model T was being converted into basic sports, or even racing, cars to provide rivalry to a new breed of car which had captured the American motorists' imagination.

These were the 'raceabouts' – as exemplified by the Stutz and Mercer models of the 1910s. Their formula for success was to look good, to have bright colors, and big engines in very simple, light, chassis. Built only as two-seaters, they featured minimal weather protection, and had exposed fuel tanks behind the seats; this was supposed to enhance the weight distribution, and improve safety.

Yet by the 1920s, cars like these were barely acceptable, in a country where large engines were then commonplace, but where enclosed bodywork was coming into vogue. Because they usually had great distances to travel, the Americans searched for a more practical way of motoring from A to B, and showed less interest in sporting cars of the types well-proven in the early Glidden Tours.

Meanwhile, in Europe, a more advanced type of sportscar was evolving, a type exemplified by the Austro-Daimler, the Prince Henry Vauxhall, and the Alfonso Hispano-Suiza. These were all scientifically-designed machines, developed to give both reasonable performance, good road-holding and handling. Controls worked with precision, and the weight distribution was good.

Naturally the four years of war, from 1914 to 1918, stifled car development for a time, but indirectly it was to lead to a revolution in the building of high-performance cars. W O Bentley, in Britain and Marc Birkigt, in France and Spain, were among numerous talented designers who had been involved with aero-engines during the war. The lessons learned were then applied to the new breed of more sophisticated car, such as the Bentley and the Hispano-Suiza. Almost invariably, engines had overhead camshaft valve gear, along with aluminum castings for many mechanical components.

Four wheel brakes, which followed, owed much of their technical success to the techniques honed in war, even though they had not been developed for aircraft. Front-wheel brakes, in fact, had been tried out at least five years before the war, but had not been very successful, due to the problems of providing compensation between front and rear.

Straight after the war, however, Duesenberg and Hispano-Suiza came up with very practical systems which at last allowed their engines' power to be used to the full, in safety on the road. Another important wartime development for aircraft had been the use of superchargers for engines, and as the 1920s progressed these were to become an important means of boosting power on sports and racing cars.

After the war, in Europe and in America, large manufacturers found ways of offering more satisfactory cars made cheaper by mass-production techniques. Morris showed the way, with Austin, Citroen, Fiat and others all following suit. The younger, livelier, motoring element sought more exciting ways, and more interesting machines, than these, and either built their own sporting cycle-cars from existing components, or else bought complete cars from companies like Morgan, GN, or other light car makers.

It did not take enthusiasts long to realize that mass-produced family cars could be a splendid source of useful components, or as a sound basis for tuning. The Austin Seven decimated the cycle cars in the 1920s, for it was soon found to be ripe for 'customizing.' Even more significant than this was what Cecil Kimber's Morris Garages concern could do to the 'bullnose' Morris cars they sold.

To begin with, there was little more than a smart paint scheme, a few accessories, and an MG badge. Later, more serious mechanical improvements were made, and from about 1930 these slowly began to influence the family cars on which they were based. Brakes and tires improved, as did engines, with better power, efficiency, and – indirectly – fuel consumption.

The 1920s saw the inception of the Le Mans 24 Hour race, which became a real showcase for sportscar manufacturers to

Right: 1924 Bugatti T35.

BUGATTI

make their point. Bentley, for instance, won at Le Mans in 1924, 1927, 1928, 1929, and 1930, although they never capitalized on this domination to reap any commercial success.

Many true production-type sportscars followed in the wake of MG, and it was marques like those which spelt the virtual end of the French sportscar industry, where firms like Derby and Salmson had flourished by making better cars than the cycle cars. Bugatti, however, was out on its own, and quite immune from such pressures. Like Bentley, Bugatti made a series of truly great sportscars, but it also failed to make the most of the many successes. This was largely because of the need to charge a very high price, for a car which made no compromises in the materials used, or in its design and construction. In America, true sportscars had virtually died out by this time, although roadster versions of all sorts of popular cars were still available. True specialists like Duesenberg and Stutz had moved up-market, and all suffered badly following the Wall Street Crash of 1929, by which time their luxurious high-performance cars could be afforded by very few customers.

The 1930s opened with ever-cheaper sportscars like the MG Midget evolving from the mass-produced family cars (in that case, the Morris Minor). Spartan two-seaters still had a following, but there were signs of a move toward cars with softer springing, and more luxurious fast tourers. This type was exemplified by the 3½-liter Bentley – the 'Silent Sports Car,' which was actually built by Rolls-Royce after that revered company had taken over the equally respected Bentley concern. W O Bentley himself eventually moved on to Lagonda, where he worked on more of the same sort of semi-sporting, but fast, glamorous cars, which were also produced by companies like Alvis.

A cheap way of building cars to this formula had been discovered by Railton. This company made good-looking cars in the traditional British style, but got round the expensive problem of equipping them with sophisticated engines, by installing rather basic, but nonetheless powerful and reliable American engines.

Among several others to follow this course were to be Jensen, Brough Superior, Lammas-Graham and Allard. In France several small firms also adopted large and powerful engines from America in their 1930s models, but with little sales success. The Fascist governments of Italy and Germany did much to promote the sporting image of their motor industries, and produced some very fast, exciting, and flamboyant models.

In Britain on the other hand, Morgan became a sportscar force to be reckoned with when it moved up from three to four-wheelers, while Singer tried to take a leaf out of MG's book. The really tough rough, rugged and spartan sportscar was on its way out, a fact mourned by the Vintage Sports Car Club, which was founded in 1934 to ensure that such older types survived, to be loved, raced and rallied. The VSCC still thrives today.

In the early 1940s there was no time to be making sportscars, and after the World War II the world was crying out for basic saloons to make up the worn-down stock. In Britain, at this time, which was still the center of the sportscar industry, it became clear that few were willing to put up with the primitive pleasures of ownership when conventional mass-produced saloons had become so functional. However, at the international level, Britain could not hope to keep pace with the Americans, so several new types of cars were evolved.

At one end of the spectrum there was the new Land-Rover, but at the other there was a rush to produce new sportscars. Though MG and Morgan retained their pre-war looks for a time, newcomers like the Triumph TR2, the Jowett Jupiter, the Austin-Healey 100, and the Jaguar XK120 all went for international streamlined good looks, using engines from a saloon car in the maker's existing range.

The Jaguar was certainly the trendsetter, for it gave appreciably more performance than any other car in series production at the time. The more down-to-earth Triumph TRs and Austin-Healeys were good for at least 100mph, and – like MG – they found a far larger market in the USA than in their homeland.

Sportscars were useful mobile 'test-beds' for new refinements, which would later spread to saloons as speeds crept ever higher. Jaguar's racing sportscars of 1952 were the first to have disk brakes, and later in the same decade radial ply tires came along to revolutionize roadholding on all cars. Jaguar, with its famous twin overhead camshaft XK engines, came to dominate sportscar racing, and to follow in Bentley's footsteps at Le Mans.

By the mid-1950s, the success of all the European imports from Alfa-Romeo and Mercedes-Benz, had convinced the Americans to have a go themselves. First Chevrolet, and then Ford, produced their own variations on the theme – which was to have a sleek open car, with a big engine from a conventional saloon shoe-horned in. After the Corvette came the Thunderbird.

Cars of the same caliber as Bugatti (which never really got going again after World War II), were developed by Ferrari, Maserati and Aston Martin. All found, however, that the demand for expensive high-performance sportscars was strictly limited, and found it better to concentrate on the building of luxurious Grand Touring cars instead. GTs also appeared from companies like Jensen, de Tomaso, Monteverdi, Facel Vega, and Iso, all of whom used powerful American V8 engines. Jensen attempted to move into the true sportscar market, in 1972, with the Jensen-Healey, but this project failed.

Left: 1929 Supercharged 4½-liter Bentley.

Several post-war front-runners built up their image in motor sport. They also did much to encourage private entrants, so that amateur racing with true road cars continued. Relatively minor, but still influential, events were staged all over America, which did much to foster the cult.

Even if the typical early post-war sportscar was more refined than ever, with easy-to-fit hoods and sidescreens (even wind-up windows in some cases), the typical model still had a conventional chassis, with a 'cart-sprung' rear end, and other traditional features. In 1953, however, the AC Ace, which evolved from one of John Tojeiro's racing designs, marked one of the first serious attempts to make an advanced sportscar which owed little to everyday saloon car design. It featured a tubular chassis frame, with a body of aluminum panels, and had independent suspension to all four wheels.

Monocoque construction had come into widespread use in saloon cars from the 1930s, but it was difficult to make a sportscar like this, because it lacked a roof, so space frames or deep backbone chassis were tried instead. Lotus used both types and, like Porsche, was one of the great post-war success stories.

The Lotus story began following one man, Colin Chapman, being obsessed with motor sport. The Porsche, though, began as a most unlikely sportscar, for it borrowed its mechanical elements from the ultra-proletarian Volkswagen Beetle, and was therefore condemned to having a rear-engined layout. The afficionados said that no self-respecting sportscar could possibly have an engine in this position, and handle properly. This was true of early Porsches, but later models improved, and many race and rally successes finally confounded the critics.

Lotus, like other specialist producers of sportscars, such as Alpine-Renault, TVR and Reliant, all found that their small output did not justify the high-tooling costs of metal body pressings, and made use of the new 'wonder' material, fiberglass. So, too, did the Chevrolet Corvette, in America. In the case of the Lotus Elite, which first appeared in 1957, there was a complete monocoque of fiberglass.

In America, at this time, there were different trends, as public interest in drag racing, and stock car events, led to the development of two new types of cars – the 'pony cars,' of which the Ford Mustang and the Chevrolet Camaro were perfect examples, and the 'muscle cars' such as the Pontiac GTO and the Plymouth Road Runner. Ford and Chrysler (but not GM, officially), put much money and effort into motor sport.

In Europe, the gulf between racing and roadgoing sportscars widened in the 1960s, with mid-engined types coming to dominate the former. Front-engined road cars gradually fell back, in terms of technical sophistication. For many manufacturers, the truth was that profits and sales no longer justified a big expenditure on development, so cars like the Austin-Healey 3000 and the Sprite, the Triumph Spitfire and the TRs, the MG Midget and the MGB simply soldiered on, and on, becoming anachronisms. Even though the Jaguar E-Type of 1961 was both advanced and refined, it was eventually replaced by a GT car in the mid-1970s.

AC's attempts to get back into the sportscar market after the macho Cobra had died, and the mid-engined car sold very slowly. Lotus, for their part, moved up-market as their small open cars became harder to sell. Morgan stuck to its traditional formula (and built nine cars a week!) with great success, relying in later years for considerable help from the modern lightweight Rover V8 engine. This engine went into several other cars, including the TVR in the 1980s, giving that firm a new lease of life.

Many marques have disappeared in recent years, for only Matra and Alpine-Renault survive in France, but in Germany Porsche is now completely unchallenged, as Mercedes-Benz have gradually turned over to building increasingly refined, and less sporting, open tourers.

Most of West Germany's efforts have been channeled into producing sports saloons, of the type made popular by BMW, a firm which in the 1930s was renowned for making sportscars. Nowadays, the BMW saloons, and other impressive performers such as the VW Golf GTI, can do everything that the traditional sportscar can do, and all this in considerably greater comfort.

In Italy, Alfa Romeo and Lancia have also gone for this market, while Fiat has pulled out to leave the market clear to coachbuilding firms like Bertone and Pininfarina. In Sweden Saab has produced a series of very popular turbocharged models.

Cars like these, the MG Metro Turbo, and the Maestro and Montego sports saloons are all outside the scope of this book, because their original concept was as very ordinary saloons or hatchbacks. Maybe a car like the Lotus-engined Sunbeam hatchback comes close to bridging the gap, and we have actually included the Audi Quattro as a sportscar because of its advanced chassis, including four-wheel-drive. The mere fact that a car's bodywork is totally enclosed does not damage its status as a sportscar, as witness most Porsches ever made. However, few enthusiasts would deny that their conception of a sportscar is a low-built, open two-seater (or perhaps a 2+2), with masses of bonnet in relation to passenger or luggage space. It probably has an uncompromisingly firm ride, very direct steering, and an exhilirating exhaust note.

Unfortunately, such features inevitably led to decline of the breed. The problem was that most sportscar manufacturers had to rely, almost totally, on the American market, which suddenly became safety-conscious, and emissions-conscious during the 1970s. Open cars without safety roll-

Right: 1932 Alfa Romeo 8C, 2300cc Le Mans.

Alfa Romeo

over bars or cages began to be viewed with great suspicion, and by the time energy absorbing bumpers had been fitted to cars like the MGB, much of their styling appeal had been lost. A mandatory crash-testing program was beyond the resources of most small manufacturers, and to make engine and exhaust system modifications to meet emission limits robbed many an engine of its crispness.

One by one, therefore, British and Italian manufacturers abandoned America, and the market was left wide open for one brave manufacturer with big resources. That manufacturer was Nissan (Datsun) of Japan.

At the end of the 1960s, this company launched the large Z-Series sports coupes, which had much of the character of the old Austin-Healeys, and they soon became the best-selling sportscars of all time, with three quarters of a million delivered by the end of 1980.

These cars unashamedly borrowed ideas, and looks, from European and American cars, and to many purists this made them an unacceptable compromise. Yet the Z-cars were fast, and good fun, and Americans in particular took to them whole-heartedly. Other attempts to build sportscars, by Japanese manufacturers, were not as successful, though the Wankel-engined Mazdas, and the little Honda were technically interesting. The Honda S800 was the motorcycle giant's first attempt to produce a car, and could manage 100mph from a 791cc engine, but Europe and America were not used to the idea of tiny hard-working engines, and the project fizzled out in 1969.

We have already noted that there was a trend toward mid-engined high-performance cars in the 1960s. The Ferrari Dino first appeared, as a prototype, in 1965, while Maserati and Lamborghini soon favored mid engines, to give optimum roadholding balance. Lotus and others followed suit, and many felt that the Rover P6BS prototype, might have been a winner if put into production.

As it was, most of the exotic cars used a layout which was neither very practical, nor offered any sensible luggage space. Most were very individual 'dream cars,' which were seriously affected by the energy crisis of the 1970s.

It was at about the same time that the last of the truly sporting Aston Martins was built – the twin-cam 325bhp DB6 – to be replaced by more luxurious and refined front-engined Grand Touring cars. Virtually all the hybrid American-engined European marques disappeared, with the honorable exception of the Bristol, which was in any case more of a gentleman's sporting carriage than a true sportscar.

There was one encouraging revival at Bentley, which had for years merely been building badge-engineered Rolls-Royces. The marque took on a more sporting image in the early 1980s, with the Mulsanne Turbo, which was a justifiable throwback to the legendary 'blower' models of the 1920s and 1930s.

With most of the lucrative American market now dominated by Datsun, and by the large corporations, there seemed to be little hope for newcomers. In the 1960s, however, there was a surprising development which soon became quite successful. The car which started the trend was the Excalibur, from Wisconsin, in 1964. This was a near-lookalike of the 1920s-style Mercedes-Benz SSK, but which used all the latest mass-produced American mechanical items. Many other American firms subsequently entered this 'replicar' field, making copies of Duesenbergs, Cords and Auburns, along with their own ideas of what they thought a 1930s-type 'Classic' should look like (sometimes with distinctly unaesthetic consequences!). This idea later spread to Europe, where Bugattis, BMWs, Alfa Romeos, Delahayes, Delages, Jaguars, MGs and even Ford GT40s were soon being copied, at least as to their outward appearance.

In Britain, at least, alongside these expensive hand-made models, has grown up a sizeable industry, that relies on supplying kits for home assembly, often around secondhand mechanical components. Such cars can often be created from saloons which have failed their roadworthiness tests. After the bodywork has been discarded, and the components used in the new kit, something looking much like an open Morgan, or an MG, can be created at home by anyone with average mechanical skills and a good assortment of tools.

In the mid-1980s, however, the classic British sportscar is virtually extinct, apart from small concerns like Morgan. Meanwhile tens of thousands of obsolete models are now lovingly preserved by enthusiasts around the world. Sports coupes of the sort made by Porsche and Datsun have swept the board, although there is now a trend back toward open cabriolet versions of the 'hot' saloons. Among the surviving makers, TVR and Reliant returned to offering soft-top versions, and now it looks as if this trend will continue. The arts of turbocharging and fuel injection have improved engines greatly, these help to clean up exhaust emissions and to make sportscars more acceptable in environmental terms.

The 'replicar' market continues to grow, as does the 'kit car' business. Perhaps there is something to commend in this, for many people feel that it is ecologically unwise to dispose of 'energy intensive' metal components if some of them still have a useful life. An enthusiast willing to strip down engines, axles and transmissions in his own time, is not constrained by financial considerations, and might end up with a really distinctive and individual sportscar that might have cost tens of thousands of pounds to produce by traditional means.

Left: 1936 Mercedes-Benz 540K Cabriolet.

AC Cobra

The Cobra came into existence in 1962, as a comprehensive redesign of the AC Ace, which had originally been announced by the Thames Ditton-based concern, in England, in 1953.

The Ace, for its part, was a lightly-modified version of the Tojiero sports racing chassis, which had a 'ladder-style' chassis frame, with three inch diameter tubular side members, and independent suspension all round, by transverse leaf springs and wishbones. The original power unit was AC's own light-alloy 2-liter six-cylinder engine, but by 1955 the higher performance six-cylinder Bristol engine was also available. This was taken out of production in 1961, and in its place AC offered tuned-up versions of the British Ford Zephyr 2.6-liter six; the origins of this engine were really too mundane to appeal to buyers of such a specialized sportscar.

In 1961, the Texan Caroll Shelby, a retired racing driver whose successes included a Le Mans victory for Aston Martin in 1959, decided that a modern, lightweight, American V8 unit would give an Ace quite startling performance. He approached AC's Hurlock brothers for support, and found them enthusiastic. The first prototype was built at Thames Ditton then shipped, engineless, to Shelby's workshops in California for further development.

By this time Ford of Detroit's interest in high performance motoring had been rekindled, and one result was that they offered Shelby supplies of their 'small-block' engines, along with development assistance, this encouraging him to start racing and selling the cars.

The first 75 production cars, now named Cobra, had the Ford 260ci (4.2-liter) engine, after which the famous 289ci (4727cc) unit was fitted instead. In standard form it provided 271hp, but it was possible to buy a race-tuned version, offering a mighty 370hp.

Right from the start, AC supplied cars to California without engines and transmissions, which were fitted by Shelby American. The cars were then tested, before delivery to customers.

In the first three years, many development changes were made to the Cobra, the most important of which being the fitting of accurate rack and pinion steering (after the first 125 cars had been built), and during 1965 the old-style transverse-leaf spring suspension was replaced by a more modern coil spring and wishbones installation, to which Ford gave some design assistance.

Later in that year, the mighty 427 Cobra was put on sale, a car in which the main chassis side members were of four inch diameter steel tubing, the engine was the massive 'large-block' 427ci (7-liter) Ford Galaxy-type. There were larger tires and more bulbous wings to cover them, and a modified nose to improve radiator cooling.

In certain classes of racing, the Cobra (particularly the 427 version) was very successful, for it was up to 1000lb lighter than its principal rival, the Chevrolet Corvette Sting Ray. Chevrolet would dearly have loved to produce a Cobra beater, but management stopped the only likely candidate, the GS Model.

Until the very specialized GT40 was race ready, the AC Cobra, sometimes sneakily called the 'Ford Cobra' by unscrupulous publicists, was often used in international sportscar races, but it was never sophisticated enough to take on all comers. In 1965, however, the Cobra managed to win the World GT Championship, against rather 'thin' opposition, not in its standard form, but with the brutal but efficient Daytona Coupe style, which had been evolved for Shelby by Pete Brock.

By the late 1960s, the Cobra was past its heyday, and there was even space at Thames Ditton to make Europeanized versions of the 4.7-liter engined car, which they called AC 289 (no mention of the

Cobra, please note). The final Cobra, for the USA, was built in 1968.

There was a sequel. To keep the pedigree alive, AC built a long-wheelbase version of the chassis, installed the more 'soft' and docile 428ci Ford V8 engine (of 7014cc), sent them to Frua of Italy, the body builders, and marketed them as AC 428s, until 1973.

In recent years, many look-alike Cobra replicas have been put on sale, some mechanically similar, and some very different under the skin. Some, indeed, have had fiberglass bodies and quite different engines, though most of them use readily-available Ford V8 engines of 428ci or 429ci.

One 'replica,' however, is exact, the reason being that it uses the remaining tooling and equipment. This is the Cobra 'Mk IV' built by Auto Kraft Ltd, whose premises are inside the confines of the old Brooklands race track, at Weybridge in Surrey. Their original business was in repairing and restoring Cobras, but they have now been granted the use of the 'Cobra' trademark and are supplying cars to Ford dealers across the United States. The latest cars incorporate engines meeting all current emission requirements, and are 305ci/5-liter units, as used in the Ford Mustang, and styling is identical with that of a late-model Cobra 427.

All pictures: 1965 AC Cobra, in road trim (above right and below left), and ready to race (below right).

Alfa Romeo

The origins of Alfa Romeo can be traced back to the Societa Anonima Italiana Darracq (SAID), which was set up to import French Darracq cars into Italy. Although this marque was very successful in France, it made no impact at all in Italy, so SAID's managing director, Ugo Stella, decided to change his strategy. Having engaged an established car designer, Giuseppe Merosi, he set him to produce new machines, ditched the agreement with Darracq, and started up again in Milan as the Societa Anonima Lombarda Fabbrica Automobile. It was not long before that cumbersome title, logically enough, was shortened to ALFA. It was not until 1915 that Nicola Romeo, already a successful businessman, moved in, took control of the business, and allowed the famous name of Alfa Romeo to emerge.

The earliest sporting Alfa Romeos to make their name were the 22/90s, or RLSS models, where 'SS' meant Super Sport. Announced in 1925, this was a good-looking car with a pronounced V-radiator; 392 were sold in two years. There were three different body styles, the most striking being a two-seater with a long tapering tail.

Performance was good, but not sensational – the 3-liter engine helped provide a top speed of at least 80mph. Peak power was 83hp, and the best cars were those produced with close ratio transmission. In some ways these cars were still mechanically crude, for there was no pressure oil feed to the engine valve gear, which had to be hand lubricated every 1000 miles.

In 1927 the first of a new six-cylinder generation arrived, this being the 1.5-liter Turismo model, which had a single overhead camshaft engine. Next along was the Gran Turismo, even more sporting, and this time with an advanced twin overhead valve cylinder head; the most sporting versions of all were equipped with Roots-type superchargers.

In 1929, however, the twin-cam engine was enlarged, to 1752cc, though the same basic chassis and running gear was retained. Here was the birth of the famous '1750' model, which is now acknowledged as a masterpiece. Designed by Vittorio Jano – an engineer still producing famous sports and racing cars 30 years later, it caught the imagination of the motoring world, and went on to become a famous Alfa Romeo competition car.

The 1750's engine had considerably more power and torque than the 1500, and

Right: A 1926 Alfa Romeo RLSS drophead coupe.
Below: 1931 Alfa Romeo 1750 GS.

gave the car better acceleration and flexibility. The gearchange was extremely positive, the steering precise, and the roadholding excellent. Of the various types of 1750, the Gran Sport and the Super Sport were the most exciting, and these were often provided with elegant, attractive body styles by coachbuilders like Zagato and Touring.

One measure of the 1750's capabilities was proved in the 1000-mile Italian Mille Miglia of 1930, which Tazio Nuvolari's 'blown' 1750 won outright, beating such cars as the 7-litre/225hp supercharged Mercedes-Benz SSK by more than one hour.

Inevitably, more power was needed for the next generation of cars, so Jano was encouraged to produce a new engine, choosing a mighty straight eight-cylinder unit, of 2336cc, using the same bore and stroke dimensions as the 1752cc six. Like the six, the new eight had twin overhead camshafts, but to overcome the torsional vibration, and strength, problems of using such a long engine, he arranged it, effectively, as two four-cylinder engines, back to back. There were separate cylinder blocks on a common crankcase, and the crank ran in no fewer than 10 main bearings to keep it as rigid as possible. As before, there was a Roots-type supercharger, mechanically driven by gears from the center of the crankshaft, drawing fuel/air mixture through a Memini carburetor, and the engine produced no less than 138hp at 5000rpm.

The chassis was still flexible, but at least this was balanced by the use of rock hard front and rear leaf springs. At the front axle, there was extra location by radius arms, and in true Alfa Romeo style the steering was quite delightfully light, while the roadholding was exciting, but predictable. This new car was known as the 8C 2300, and was offered either as the Lungo, or the long-wheelbase (10ft 4in) Le Mans, or as the Corto, or Mille Miglia, with a wheelbase of only 9ft 0in. Compared with the 1750, the braking was much improved (for very large finned alloy brake drums were fitted), but as usual the clutch operation was very fierce, and needed to be mastered by an expert.

Competition successes came straight away in 1931, when the car was put on sale, with outright wins being notched up in the Targa Florio and the Le Mans 24 Hour race that year. Two famous versions to evolve from this basic design were the 8C 2300

Above: 1938 Alfa Romeo 8C 2900 B Touring.
Overleaf: 1932 Alfa Romeo 8C 2300 Monza.

Monza, an open-wheeler two-seater racing car, the 2556cc derivative of that car, and of course the Tipo B (sometimes called P3) Grand Prix car of 1932, which used the same basic supercharged eight-cylinder engine.

In 1936, a new model, the 8C 2900 appeared, which used a 2905cc engine, developed from those fabulously successful P3 single-seaters, this being the largest practical 'stretch' of Jano's celebrated straight-eight unit. This car was produced in two versions – suffix -A for competition use, and suffix-B for touring, this type being produced with short-wheelbase, or long-wheelbase chassis, and a wide choice of body styles. In this form, detuned from the racing cars, the 2.9 had twin superchargers, and twin Weber carburetors, producing an astonishing 180hp at 5000rpm. Other modern features included the use of hydraulic brakes, hydraulic front dampers and independent front suspension by coil springs.

The 8C 2900s were the largest and the most sophisticated Alfa Romeos yet put on sale, but they were so expensive that only 30 were sold in two years. However, by this time Alfa Romeo had fallen into the hands of the Italian state, through its IRI financial conglomerate, and was busily building weapons of war, so private car production was not thought to be important. During World War II the Milan factory was devastated by bombing, so it was not until 1950 that an all-new saloon car, the 1900, could be announced.

Although the 1900 sold well by any previous Alfa standards, the company set out to expand even further by designing a small-engined range, the first of which types, the Giulietta Sprint, was seen in 1954.

Every car in the Giulietta family had the

Alfa Romeo
DYE 484

Above: 1954 Alfa Romeo Giulietta Spider.

same type of 1290cc twin-cam four-cylinder engine, of almost 'traditional' Alfa Romeo layout. Bore and stroke dimensions were nearly the same, breathing was efficient, and the engine revved easily to 7000rpm. Instead of a separate chassis, the Giuliettas were all based on a unit construction body/chassis design. There was coil spring and wishbone independent front suspension, with an anti-roll bar, while the rear axle, though sprung on coils, had a complicated location which included trailing arms and an A-bracket linking the floor pan to the differential casing.

The Sprint, actually announced before the saloon, was the 'bread and butter' model in the range. It had a beautiful body styled by Bertone, offering four seats under the coupe roof. The engine produced 80hp, and top speed was almost 100mph. A year later, the Sprint was joined by the Pininfarina-styled open two-seater Spider, which had a five-inch shorter wheelbase version of the same floorpan, but the same running gear.

A year later higher-performance (90hp) versions of these cars were announced, and given the model names Sprint Veloce and Spider Veloce; both were capable of 110mph. Later in the 1950s, two extra special versions of this 'chassis' were put on sale – one being the Sprint Speciale, which had a stylish Bertone body with lengthened nose and tail, the other being the Sprint Zagato, which had a wickedly stubby little coupe style by Zagato. Both these cars had 100hp engines, five-speed transmission, could reach 60mph in less than ten seconds, and had top speeds of around 120mph.

In 1962, the Giulietta was progressively succeeded by the Giulia range, which used much the same running gear and suspension, but which had an engine 'stretched' to 1570cc. The saloon, called Giulia TI, had a boxy style, a single carburetor version of the engine, 92hp, and a top speed of around 100mph. At first there were interim sporting models – both the Giuliettas getting the larger engine and becoming Giulia Sprint and Giulia Spider respectively.

The definitive sporting Giulias, however, were the Sprint GT of 1963 (with 106hp, five-speed transmission and luscious styling by Bertone), and the controversially-styled 1600 Duetto Spider (with the same running gear, and Pininfarina bodywork). Over the years, many derivatives of these two styles were produced, not least the lightweight Giulia GTA, the 115hp 'homologation special.' The Duetto's styling was made more conventional for 1970, when the tail was cut short, by which time the car had come to be known, simply, as 'Spider'.

Below: 1963 Zagato-bodied Alfa Romeo Giulia T2 1600.

Above: 1954 Alfa Romeo Giulietta Sprint 1300.

Over the years, engines between 1290cc and 1962cc, all of the same twin-cam family, were fitted to these cars, which were marketed in a bewildering number of guises, the Spider still being sold in the USA in the mid-1980s.

The most remarkable Giulia of all was the GTZ (Z = Zagato) Tubulare, which had an entirely special multi-tube space frame chassis, and light-alloy body style, but the familiar engine. This was strictly a racing car, not intended for normal road use – although some of them unaccountably found their way onto the public highway.

At the Montreal exhibition, Expo 67, Alfa Romeo showed a Bertone-styled 'special,' which they called the Montreal, and public interest was such that they put the car on sale in 1970. Based on the Giulia GTV's floorpan and suspensions, the Montreal was powered by a de-rated, 200hp, version of the four-cam V8 engine already used in the Type 33 racing sportscars, and in this form it had 2593cc with fuel injection. Built in limited numbers until 1977, it was available with right-hand-drive from 1974. In standard form, the Montreal had a top speed of 136mph.

Below: 1964 Alfa Romeo Giulia GTC Spider with 1.6-liter twin-cam engine.

Allard

Sydney Allard was a motor trader from London who liked to go racing, and to compete in trials. In 1936 he built the very first Allard Special, which was based on a Ford V8 chassis and running gear, but which also used part of a GP Bugatti's body shell! He then went on to build another 11 Specials before the outbreak of war, all Ford based, with V8, or Lincoln Zephyr V12 engines.

After the war, while the family retained its Ford dealership, Allard set up the Allard Motor Co., to build cars, announcing his first models in 1946. These had side-valve Ford V8 engines and transmissions, special chassis, and split-beam independent front suspension with transverse leaf springs.

The original competition two-seater was the J1, and this was replaced by the legendary J2 in 1950; this car had an aluminum body, separate cycle-type wings, and a ladder-style frame, while there was now coil spring independent front suspension, and a De Dion rear end. This car had a top speed of 110mph with 160hp Cadillac V8s power (a whole variety of engines could be fitted), though most of the cars sold in Britain had ex-military 4.4-liter Mercury V8s, some with Ardun overhead-valve conversion cylinder heads, which had been designed by Zora Arkus Duntov.

It was with the Cadillac engine fitted that Allard himself, co-driven by Tom Cole, took third place at Le Mans in 1950, and Allards of this, and the later J2X type, won hundreds of races in North America during the 1950s. 'X,' in this context, referred to the new design of tubular chassis frame applied to all Allard cars from 1952/53 onward. It was probably Allard, incidentally, who started the tradition of exporting cars to the USA without engines or transmissions, for enthusiasts to fit themselves.

From 1953 (in spite of a Monte Carlo win for the intrepid Sydney Allard in 1952), Allard's sales fell away as export competition from cars like the Jaguar XK120s became too fierce. The less sporting models, such as K, M and P types were not fast enough, and the K3 Tourer was not sporting enough. Allard tried to boost sales by going 'down-market,' with the Palm Beach models, which were smaller, smoother-styled, and with British Ford four-cylinder or six-cylinder engines, but these were commercial failures. So, too, was an attempt to sell Allards with twin-cam Jaguar engines, and the last cars of all were built in 1959.

Allard himself was a great European drag racing enthusiast, and later dabbled successfully in the tuning-up of small Fords. He died in 1966.

Since 1981, a modern 'replica' of the J2X has been on sale, manufactured in Canada, using Chrysler V8 engines.

Below: 1948 Allard K1.
Right: 1948 Allard K1 racer.

Below right: 1952 Allard K2, superseded by the K3 in 1953.
Far right: 1952 Allard K2 Roadster.

346 JGX
152E

WHX 77
17

Alpine Renault

Over the years, many small specialist concerns have sprung up, which concentrate their attention on the products of one major motorcar manufacturer. Although there may not be any financial involvement, such companies often receive technical support and assistance, and both parties benefit. One such specialist is Alpine of Dieppe, who have enjoyed a continuous relationship with Renault since the 1950s.

Jean Redélé was a young Renault dealer, who built his first racing 4CV saloon in the early 1950s, entered it successfully in the Mille Miglia, and went on to produce for sale the Type A106 Alpine-Renault in 1954. This light, rear-engined two-seater had a 4CV platform, but a fiberglass body, along with the suspensions and power unit of the Type 1063 Renault saloon. Optional were the 747cc 'normal' or 48hp 'Mille Miles' models.

To supplement, and eventually replace the A106, came the A108 of 1957, which had a new steel backbone frame chassis, 4CV/Dauphine type of suspension, and 845, 904 or 997cc Renault engines, hung at the rear as usual. The 'classic' Alpine style introduced at this point persisted, in convertible or coupe form, into the mid-1970s, by which time the car had become Type A110, and entirely different Renault engines of up to 1647cc and 127hp. For a short time in the 1960s there was also a rather ugly four-seater, but the vast majority of all these cars were the famous low Berlinettes.

With Renault encouragement, examples were entered at Le Mans and other French races, but the cars really came to be famous in rallying in the 1960s and early 1970s, where their combination of traction (due to a rear-biased weight distribution), and 'nervous' handling made them very versatile. The rallying effort, with fleets of French blue Berlinettes came to a triumphant conclusion in 1973 when Alpine-Renault won the World Rally Championship, sometimes using up to 1.8 liters and 180hp.

For many years, the A108/A110 models had coil spring independent front suspension, with rack and pinion steering, and swing axle rear suspension, also by coil springs, but from 1973 the more sophisticated wishbone rear end of the larger Type A310 was adopted.

The larger Alpine-Renault, the A310 coupe, was launched in 1971. Although it retained the same basic chassis and mechanical layout (with the 1605cc engine from the Renault 16TS), it was clothed in a smoother fiberglass body, which had been styled by Fiore. At first the power output was 127hp, and with a five-speed transmission the stop speed was more than 130mph; it was also possible to order a 1.8-liter 'Tour de France' engine with 165hp.

From the autumn of 1976, the A310 was offered with the PRV 2664cc, V6 engine, which had 150hp, and much more torque than the four. This version was, logically enough, called the A310-V6 and could reach 137mph. Eventually, it took over completely from the four-cylinder car, and was replaced by a different style of Alpine-Renault – the V6 GT and the V6 Turbo. Closely based on the long-established chassis engineering, and with the PRV engine still in the extreme tail, the latest car had a longer wheelbase, and wider tracks. The normally-aspirated engine produced 160hp from 2849cc, while the turbocharged car had 200hp from only 2458cc. Top speed of the Turbo Coupe was predicted at more than 155mph. Unlike the previous model, however, this Alpine-Renault had been designed completely by Renault themselves, although assembly was still to be at the well-known Dieppe factory, owned by Renault for the past decade or so.

Top right: 1300cc Alpine-Renault on the Targa Florio.
Right: 1972 Alpine-Renault 1600S on the Acropolis Rally.
Below: 1981 Renault Alpine A310 1600 Coupe.

118
118

Alvis

The Alvis car was made in Coventry, England, by a company founded by T G John to produce automobile engines, castings for carburetors, and motor scooters. John himself was originally a naval architect, who became chief engineer of Siddeley-Deasy, in 1915, where aeroplane engines were to be made.

One of his associates was G P H de Freville (who had imported DFP cars to Britain before W O Bentley had taken on that concession). Not only did he design the engine for the very first Alvis car, the 10/30 of 1920, but he also invented the marque name Alvis. This had no significant meaning, other than it rolled nicely off the tongue!

Right away sporting Alvis cars were produced, for the 10/30 model could reach 60mph using only a 1.5-liter engine, which was superior to most of its rivals. A Super Sports version was well received, and this featured the 'duck's back' body style for which Alvis cars later became famous.

In 1923 the 12/50 model was announced, and this achieved instant fame when an

Right: 1934 Speed 20 tourer.
Far right: 1930 Alvis 1500 front-wheel-drive.
Below: Still competing in historic events, an Alvis 4.3-liter.
Bottom: 1923 Alvis 12/50 SA Duck's Back.
Bottom right: 1932 Alvis Speed 20 Vanden Plas drophead coupe.

YS 22

VC 4364

Above: 1928 Alvis 12/50 hp at St Martin's Corner, Boulogne Speed Trials, 1928.

early example won the Junior Car Club's 200 miles race at Brooklands at an astonishing 93.3mph. Road-equipped cars, even then, were capable of 70mph. The 12/50's new pushrod overhead valve engine was designed by Captain G T Smith-Clarke, who had arrived as chief engineer from Daimler, and his chief draughtsman was W M Dunn, who later succeeded him. Between them, they were responsible for all but the very last Alvis private cars.

In 1924, 600 cars were sold in the year, and the 12/50 type engine was used in every one. Thereafter the engine was continuously improved, with a guaranteed 60mph being claimed for the two-seaters by 1927. The duck's back models were still offered, but the popularity of this style fell away.

In 1925 an experimental front-wheel-drive racing Alvis was produced, where the 12/50 engine was used. This prototype set up fastest time of the day at the Shelsley Walsh hillclimb that year. Alvis became so addicted to front-wheel-drive that they built very special eight-cylinder racing models (which were not successful), and put a front-wheel-drive sportscar on sale. These were troublesome and expensive (only 155 were built) compared with 9000 conventional Alvises built up to 1932.

Even though the first six-cylinder Alvis was launched in 1928, the successful 12/50 refused to die, and though withdrawn once, it soon had to be reintroduced into the list. In 1931 a twin-carburetor version known as the 12/60 was also offered, this usually having 'beetle back' bodywork, where there was an extra 'dickey' seat hidden away in the rear bodywork. All the 12/50s and 12/60s were lively and fun to drive, strong and reliable. Many have survived to this day.

Meantime the six-cylinder Alvis line developed, with a touring Silver Eagle model being given a 2.5-liter version in

Left: 1938 Alvis 4.3-liter Vanden Plas. This dates from long after the company's abortive excursions into front-wheel-drive cars.
Right: 1950 Alvis TB14.

Above: 1963 Alvis TD 21 drophead coupe.

1931. Later in the year this engine was given to the new Speed Twenty, reputedly produced from drawing board to actuality in a mere 14 weeks, by Smith-Clarke's team. Alvis, incidentally, never built their own body shells, preferring to buy in graceful styles from companies like Vanden Plas, Charlesworth, Mayfair, Cross & Ellis, or Thrupp & Maberly.

Even as the Speed Twenty this car could reach 90mph; it was given a 2.76-liter engine in 1935, and shortly became the Speed Twenty Five when a 3.57-liter version of the same engine was installed. These were fine cars, handsome and low-slung, which could reach 95mph in ideal conditions, and were worthy successors to the four-cylinder 'vintage' Alvises.

Perhaps the most exciting 1930s Alvis of all was the 4.3-liter, which was announced in 1937. This was one of very few cars of the day capable of a genuine 100 mph, for it had 137hp in standard form, and up to 170hp could be extracted from race-prepared units.

World War II, and German bombing, not only put an end to Alvis production, but nearly destroyed the marque completely when the factory was blitzed. Postwar Alvis models were the TA14 of 1947, and the TC21 of 1950, both of which were sedate, nicely-built touring cars. However a handful of TB14 and TB21 sports tourers, with Panelcraft bodies on the standard chassis were made.

From 1950 onward, all cars were based on the TC21 chassis and 3-liter engine. With traditional, or with (from 1958) Graber styling, drop head coupe styles were available, but the days of the Alvis sportscar were over.

Alvis were taken over by Rover in 1965, and there was one intriguing 'might-have-been.' Rover designed an advanced mid-engined sports coupe, using their own light-alloy 3.5-liter V8 engine, and in 1967 Alvis were considered as candidates to produce the car. Then came the British Leyland formation, which involved Rover-Alvis, and this project was canceled.

DJA 554
4

KF 983

Aston Martin

When one considers the pure-bred pedigree of the Aston Martin of today, it is difficult to accept the fact that the first car was very definitely a 'Special.' The original car was conceived by Robert Bamford and Lionel Martin in 1913. Both were partners in a business selling Singer cars, but decided they would like to build a more sophisticated competition car.

Only the engine – a four-cylinder, side-valve, 1389cc Coventry-Simplex unit – was already in existence when they started. At the time the firm was best known for supplying power units for the inexpensive Clyno car. Lionel Martin, the engineer of the two, then fitted the engine to a 1908 Isotta-Fraschini racing chassis, which had been designed by Ettore Bugatti, no less. The car's name – Aston Martin – was derived from Martin himself, and from the name of the Aston Clinton hillclimb course, where his earlier Singer models had performed well in competition.

The first true Aston Martin production car was built in 1919, but sales began in 1921; it bore a distinct similarity to the contemporary Bugatti. These first cars retained Coventry-Simplex engines, had first class roadholding, extremely quick and precise steering and excellent two wheel (rear) brakes, although four-wheel brakes became available later.

The car had a top speed of more than 70mph, but was far too expensive, at £850, and sales were very limited. A single overhead cam engine design was a failure, and though control was bought by the Charnwood family in 1924, after which a twin-cam engine was designed, there was still no success, and the company had struck financial trouble by 1926.

The concern was then rescued by an Italian-born engineer, 'Bert' Bertelli, along with his partner W S Renwick. They went on to produce a new Aston Martin, in premises at Feltham, a conventional (by 1927 standards) two-seater, with a new 1.5-liter single overhead cam four-cylinder engine. As ever, though, the company's resources were limited, and up to 1930 only 30 cars of this type were built. There was a short spell when the Frazer Nash company guaranteed the company's overdraft with its bankers, and only a further reorganization allowed the first of the Internationals, first shown in 1928, to be on sale by 1930. For a short time W Prudeaux Brune (Aston Martin's London distributor) took control, then in 1932 Aston Martin's destiny passed into the hands of Sir Arthur Sutherland.

For its day, the International was an unusual sportscar, for it had four seats, and a low, rakish style which immediately became fashionable. The engine was much the same as earlier models, but was now fitted with dry sump lubrication. By this time the company was seriously involved in sportscar racing and trials, and were successful at Le Mans where Bertelli and Driscoll won the Biennial Cup – after which a two-seater Le Mans was introduced.

At the 1934 Olympia Show, the Aston Martins became Mk II, with a more powerful 75hp engine. One previous weak point was eliminated when the worm drive back axle was replaced by a more conventional spiral bevel type. In spite of weighing too much (about 2000lb), the car could still achieve 80mph, which was very creditable for its day.

Left: 1934 Aston Martin Ulster.
Below left: 1930 Aston Martin International.

Below: Rare 1948 Aston Martin DB1 Spa winner.

The Ulster was effectively a replica of the works racing cars, and is often regarded as one of the most attractive 1930s-type sportscars ever made. Strictly a two-seater, it had a long streamlined tail, close fitting cycle-type wings, an outside exhaust manifold, very low lines, and a most impressively equipped instrument panel. This was the best of the Bertelli Astons (for his influence was still in these cars), but it was much heavier than rivals like the Rileys, and MG Magnettes. Only 17 road cars were built, along with several 'team' racing cars, and they were priced at £750, ready to race.

In 1936, Aston Martin produced a new 2-liter model. The engine design was basically as before, but enlarged to 1949cc, and had normal wet sump lubrication. It also had a synchromesh transmission, a great advance, but ushering in a new era for the marque. There were saloon and sports tourer versions, the competition type being shown as the Speed model, which retained all the special equipment. One of these cars won the Leinster Trophy race of 1938, driven by St John Horsfall, and the same driver set best British performance at Le Mans.

The company found it difficult to get going again after World War II. Engineer Claude Hill had designed a new model, known as 'Atom,' in 1939, but a new car with multi-tube space frame and Atom's new 2-liter engine which he produced in 1946/47 could not be financed. However the industrialist David Brown stepped in to buy the company in 1947 (he also bought up Lagonda, and its splendid new six-cylinder twin-cam engine), and the new design, now called DB1, went on sale in 1948.

Above: 1935 100mph Aston Martin Ulster; this series was raced at Le Mans in 1935.

A works car, driven by St John Horsfall, won the Spa 24 Hours race in 1948, but few DB1s were ever sold. In 1949, however, the Lagonda engine, of 2580cc, was mated to a new version of the multi-tube chassis, eventually went on sale in 1950 as the DB2, and is now an all-time classic. Its aerodynamic two-seater coupe body was styled by Frank Feeley, ex-Lagonda, and sold for £1915, the drophead costing £128 more.

Serious racing began in 1950, and as a result a 123hp 'Vantage' engine (the standard unit produced 107hp) was offered, for £100 extra. A famous lightweight DB2 – VMF 63/64/65 – performed valiantly at Le Mans, where drivers Abecassis and Macklin, managed by John Wyer, won the Index of Performance; the same crew won their class in the Mille Miglia. In 1953 the DB2 gave way to the DB2/4, with 2+2 seating and a hatchback, but still in the same basic style, and from 1957, with front wheel disk brakes and yet more power, it became the DB Mk III, built until 1959.

Aston Martin also developed two special racing sportscars, using this engine. The first car was the DB3 of 1951, which had a tubular chassis frame designed by Dr Eberan von Eberhorst, de Dion rear suspension, and a 140hp version of the engine. Later cars had 2922cc engines and 163hp, a de-tuned version soon being available in the DB2/4 production car. In 1953 the DB3 gave way to the DB3S, which had a short wheelbase version of the chassis, but the same suspension and running gear, in a much more shapely body shell. This car started winning immediately – the Good-

Below: 1952 Aston Martin DB2 was the first of the DBs in series production.

wood Nine Hour race, for example – and was successful for the next three years. The DB3S was put on sale as a 'road car' for £3684, with a claimed maximum speed of 150mph. A few of these cars (there were 30 in all, including team cars) had fixed-head coupe styles.

By 1957 the DBR1, a very special racing sportscar, had been produced (it was never meant to be a road car), and in 1958 the all-new, 240hp, 3.7-liter six-cylinder DB4 was launched, as a Supercar to fight Ferrari and Maserati head on. By this time the cars were truly Grand Tourers, rather than sportscars, but the short-wheelbase DB4GT (some with lightweight Zagato bodies) produced up to 325hp, and was a very fine car. The 4-liter DB5 followed in 1963, and the full four-seater DB6 in 1965, this car being built until 1970.

The next (and most recent) Aston Martin was the DBS model, a very wide, heavy, but shapely Grand Tourer, sold at first with the 4-liter six-cylinder engine, but from 1969 with the new 5.3-liter four-cam V8 power unit. The power output of the V8 was never disclosed by the manufacturers, but the latest Vantage models were supposedly capable of 170mph.

Sir David Brown sold Aston Martin to another concern in 1972, who sold again in 1975, and the company has changed hands several times in recent years. The miracle is that through all this trauma the pedigree of the Aston Martin still shines through.

Above: 1936 Aston Martin 2-liter Speed Model.

Right: 1955 Aston Martin DB3S.
Below: 1955 DB3S cockpit.

Left: 1956 Aston Martin DB2/4 Mk11 Spyder by Touring of Milan.
Above: Spyder cockpit.
Right: 1968 Aston Martin DB6. This model saw the introduction of full four-seater bodywork.
Far right: 1970 Aston Martin DB6 Mk11 Volante.
Bottom right: Aston Martin Bulldog.
Below: 1966 Aston Martin DBSC prototype.

EGC 49H

M 88

Audi Quattro

Audi, of Germany, began building cars in 1910, and were independent for 20 years. Because of the effects of the Depression, however, they were forced to join forces with three other German car companies – DKW, Horch and Wanderer – to form Auto Union, in 1932. During the period 1934-39, rear-engined Auto Union Grand Prix cars (the first being designed by Dr Ferdinand Porsche) fought a continuous battle with Mercedes-Benz.

After World War II serious motor racing took some time to revive, and the combine concentrated on building small two-stroke engined cars. During the 1950s Daimler-Benz became major shareholders, but they in turn sold out to VW in 1964. Later the marque name 'Audi' was revived in a car partly styled and engineered by Daimler-Benz before the break, and this soon became the prestige car in VW's line-up. The first Audi 100 was launched in 1968, but the revised 100 range of 1976 included the world's first production five-cylinder petrol engine.

During the 1970s, the German army wanted a new generation of light cross-country vehicles, and VW developed the Iltis four-wheel-drive machine, which used a 1.7-liter overhead camshaft engine of VW Passat/Audi 80 type, and a modified version of the front-drive transaxle and transmission, so modified that a rear-drive propellor shaft was taken to the rear end, to drive the rear wheels as well.

One of the Audi development engineers, Walter Treser, saw the possibilities of this system in a car, installed it at first in an Audi 80 saloon and then demonstrated it to his management. Thus was the birth of the Quattro concept. Once it had been decided mate the transmission with a 2.1-liter turbocharged five-cylinder engine in a stylish coupe body, the thought of producing a rally car also took shape.

At the time (late 1970s) the world's most successful rally car was the 250hp, rear-drive, Ford Escort RS. Audi reasoned that with four-wheel-drive, and more than 300hp, their new Quattro could beat it. The car was revealed in March 1980, when it was seen to be based on a rather angular four-seater coupe body style (the Audi Coupe was launched later), in which the floor pan had been modified to allow the propellor shaft, final drive, and independent rear suspension to be installed. The engine was way ahead of the line of the front wheels, MacPherson strut front suspension, and the main transmission were that of the '200' model, and disk brakes were fitted all round. The power output was 200hp. To develop the rally car, Audi hired Hannu Mikkola, in 1980 merely on a testing contract, but for 1981 to lead the team in World Championship rallying.

The road car went on sale during summer 1980, complete with 135mph top speed, and the usual three differentials in the drive line. There was never any problem in satisfying the sporting authorities about its status, for at least 2000 Quattros were being built every year. Once established, incidentally, the 'quattro' (with a small 'q') concept was progressively applied, as an option to the other models in the Audi range.

The first 'works' rally cars produced 320hp, and were staggeringly fast, but proved to be difficult to handle. In the first season the drivers were Mikkola and the French girl, Michele Mouton, who would have won many events if it had not been for mechanical breakdowns, and accidents. In 1982, however, the marque gained seven World Championship victories, so that Audi won the World Championship for Makes, while in 1983 Hannu Mikkola himself became the World Driver's Champion.

At the end of 1983, Audi launched the Quattro Sport, which was a short-wheelbase version of the original car, with a four-valves-per-cylinder engine, and having 300hp in standard form. Just 200 of these cars were built by April 1984, for homologation purposes, but deliveries were not made until the autumn, at prices above £50,000 a car. In spite of being much quicker in a straight line (400-450hp in rallying guise) these cars were even more difficult to handle, and success was hard to achieve. Production of the 'short' Quattro occupied only a matter of months, and the original-length car continued into 1985 with refinements like anti-lock braking and an electronic dashboard display.

Below: 1983 200hp 4wd Audi Quattro turbo which was the basis of Audi 7s overwhelming success in international rallying.

Austin Healey

Donald Healey was born in Perranporth, in Cornwall, in 1898. It is truly remarkable that such a man, born and raised so far away from the industrial center of Britain, should become a successful competition driver, engineer, and eventually a car manufacturer in his own right. In 1931 he won the Monte Carlo rally outright, in a 4½-liter Invicta, and in 1933 he joined Triumph in Coventry, soon to become their technical director. After World War II, however, he set up shop in Warwick, and started building Healey cars, most of which used Riley engines.

It was in 1952 that Leonard Lord, BMC's chief executive, sponsored an informal 'design competition' for the development of a new sportscar to use Austin/BMC components. MG, even though a member of the new BMC combine, set out to produce a car (which eventually became the MGA of 1955), but it was Donald Healey's prototype 'Healey 100,' finished just before the Earls Court motor show of 1952, which took the prize.

On show opening day Lord inspected the car, offered to take it over at once, and re-named it Austin-Healey; when it went on sale in the spring of 1953 it was produced at Longbridge in Birmingham – the basic price was reduced by £100 to £1223 a few months later.

The new car had a chassis frame welded to its body shell on assembly, and used an Austin A90 Atlantic 2660cc engine, which was in easy supply following the failure of that car to sell as rapidly as hoped; the peak power output was 90hp at 4000rpm. The A90's transmission was also used, but first gear selection was blanked off, a Laycock overdrive was operative on top and second gears, the result being a five-speed transmission.

The body style was sleek and beautiful with a comfortable two-seater cockpit, and a shallow, but useful, luggage boot. There were perspex side screens, but full if rather basic weather protection, and a heater was standard. In essence, this body shell, style and chassis was to be used until the end of 1967, when the Austin-Healey 3000 finally went out of production.

Production of four-cylinder 'Big Healeys' went on to the summer of 1956, by which time 10,688 BN1s and 3,924 of the improved, four-speed transmission, BN2s had been built. Donald Healey also produced a special competition version of this car in his own Warwick workshops, of which 50 were built. Called the 100S, it had a special

Right: 1959 Austin Healey Sprite Sebring.
Below: 1958 Austin Healey 'Frogeye' Sprite, called the 'Bugeye' in the US.

Above: The Austin Healey 100-6, introduced in 1956.

Weslake cylinder head, and produced 132hp at 4700rpm, disk brakes on all four wheels, light-alloy bodywork, and no bumpers, and was an effective class-winner. Many well-respected drivers used the cars all over the world, the most startling result being the third place overall at Sebring, in the 12 Hour Race of 1954, driven by Lance Macklin and George Huntoon. A streamlined derivative of this car, with a supercharged (224hp) version of the engine, achieved 192mph on the Utah Salt Flats in 1954.

In September 1956, the redesigned car, known as the 100 Six, was launched, this having a slightly lengthened wheelbase, and the 2639cc BMC C-Series six-cylinder engine. There were modifications to the decoration, the grille, and the cockpit, but most importantly a pair of 'occasional' rear seats had been squeezed into place. The new engine produced 102hp, but the car was 400lb heavier than before, so performance, if anything, was slightly down on the BN1/BN2. When new, it was priced at £1144 in the UK, compared with – say – the £1021 asked for a Triumph TR3.

Below: 1963 Austin Healey 3000 Mk 111, the last car to bear Donald Healey's name until the Jensen Healey appeared.

In the autumn of 1957, assembly of Big Healeys was moved to the MG factory, at Abingdon, and at about the same time a more efficient, 117hp, version of the engine was produced. Later, in 1958, the two-seater body style was once more made available, and the two types carried on, side-by-side, for some years.

Finally, in mid-1959, the engine was enlarged to 2912cc, front disk brakes were standardized, and the legendary Austin-Healey 3000 was therefore born. With 124hp, it was capable of about 115mph. 3000s were built continuously for the next eight years, during which time the engine power was progressively increased to 148hp. An important re-style in 1962 gave the car a curved windscreen, winding windows in the doors, and the title of 'Convertible.' In 1964 one further change gave the car a wooden facia, and a more plushy interior. By the end of 1967, when series production ended, 57,352 six-cylinder engine cars had been produced.

The 3000 was an extremely successful competition car, particularly in rallying. Among its famous outright wins were those in the Alpine rallies of 1961 and 1962, the Austrian Alpine of 1964, and Liège-Rome-Liège of 1960 (Pat Moss) and 1964.

With BMC's encouragement, Donald Healey's designers also produced a small sportscar design, based on Austin A35 components, the result being the Sprite of 1958, which was assembled at Abingdon. Based on a simple, but strong, steel unit-construction shell, the Sprite had immense character, sufficient performance, direct steering and good handling, all at a very low initial cost. The 948cc BMC A-Series engine developed 43hp at 5200rpm, which was enough to give a top speed of about 85mph. Front suspension was from the A35, through a Morris Minor steering rack, back axle, and brakes were fitted.

The distinctive bodywork had bulbous 'frog-eye' headlamps atop the bonnet, all of which hinged up, together with the front wings, from the scuttle. There was no external access to the boot compartment – the way in being from the simply-trimmed cockpit, by folding the seats forward. At £667 when new, it sold at a very competitive price, and had no direct competition.

After 48,999 cars had been built, the 'frog-eye' was replaced by the Sprite Mk II, in mid-1961. The body shell had been redesigned considerably, with conventional squared-up front and rear styling, a normal bonnet, and a bootlid; there was also a 'badge-engineered' version of the car, built alongside it – the MG Midget. At first this car retained the 948cc engine, but in the autumn of 1962 this was replaced by a 56hp 1098cc derivative.

Then, in 1964, came the Mk III car, which had winding door windows, and half-elliptic (instead of cantilever) leaf spring rear suspension, and a more powerful 59hp engine.

The final major change came in October 1966, when the de-tuned version of the BMC Mini-Cooper S 1275cc engine, with 65hp, and a 95mph top speed, was made available. All this time, however, the MG version had been gaining in popularity, and usually outsold the Sprite every year. After the formation of British Leyland in 1968, of which Austin-Healey was a part, its days were always seen to be numbered, and the last of all was built at the end of 1970. All in all, 79,338 of the restyled Sprites had been built, and in 1971 a further 1,022, badged simply 'Austin' were also built.

This was the end of the Austin-Healey marque, but the MG Midget, of which a total of 226,526 were eventually built, stayed on sale until 1979, the last five years with a 1493cc Triumph Spitfire engine, and a 100mph top speed.

Bentley

Walter Owen Bentley was originally trained as an engineer in the Great Northern Railway workshops in Doncaster, but eventually joined the motor trade in London, and began importing French DFP cars. His first design achievement was to produce light weight, aluminum pistons for the 12/40 model, which allowed it to rev much faster, and develop more power.

Bentley then went on to become an important designer of rotary aeroplane engines for the British government during the 1914-18 war. In 1919, with the often under-estimated help of F T Burgess (who had designed the TT Humbers of 1914), and Harry Varley (ex-Vauxhall), he set out to produce his own Bentley cars, from a mews off London's Baker Street. Series production did not begin, however, until 1921, from premises which Bentley Motors took in Cricklewood, and continued until 1931.

The first Bentley production car was a 3-liter model. The engine made extensive use of aluminum in its construction, and was distinguished by the use of a single overhead camshaft, and four valves per cylinder. Peak output was about 70hp, and the car's top speed was between 70 and 80mph. At first the cars only had rear-wheel brakes, but front-wheel brakes were added in 1923, and the car's reputation was always well-founded. Every one of the famous 'WO' Bentleys – as these cars are inevitably known – was developed from this original layout.

A 4½-liter model, capable of up to 90mph, was announced in 1927, to give more power and torque. This engine produced about 100hp at first, but this was raised to 130hp in its most exciting guise for some customers. There was, however, an even more sporting 4½-liter Bentley, the famous 'Blower' model of 1929, in which the engine had been redesigned by Amherst Villiers, following encouragement from the racing driver Sir Henry Birkin. WO himself never liked the concept, but reluctantly allowed a few cars to be built, to allow it to be raced as a 'production' car. Although an unsupercharged 4½-liter model won the Le Mans 24 Hour race in 1928, the 'Blower' could never match this.

A 6½-liter six-cylinder engine, and chassis, were developed – both being extensions of the original design philosophy – and announced in 1925, to attract business away from Rolls-Royce, for the 'carriage trade.' At this stage, however, the engineering was not yet refined enough, and the project was not completely successful. However, Bentley then evolved the Speed Six sporting chassis from it, where

Right: The 'Blue Train' 6½-liter Bentley Speed Six Coupe.
Below: 1927/8 Bentley 4½-liter.

the engine developed 160hp in standard form, and the cars could cruise at 90mph. Racing versions were used by the 'works' team, winning the Le Mans classic in 1929 and again in 1930 (when a Speed Six also won the Double 12 Hour race at Brooklands too).

Bentley, incidentally, never built their own bodywork, and although this meant that some of their chassis were saddled with overweight, and – frankly – unattractive body style, it also allowed a great variety of styles to be found behind the unmistakable Bentley radiator. The most famous, of many, was the Vanden Plas open four-seater sports style, inevitably painted in British Racing Green.

In spite of all the publicity gained by Bentley, due to their many racing successes, the company always seemed to stagger from financial crisis to crisis. Woolf Barnato moved into control of the company in 1926, and also continued to be the most successful and consistent of its racing drivers, who collectively became known as the 'Bentley Boys.'

The crash came in 1931, after the magnificent 8-liter model had failed to generate enough sales to take on Rolls-Royce, and after the very 'un-Bentley' 4-liter engine had also flopped. Barnato withdrew his financial support, and the company was put into the hands of the Receiver. Eventually, after a rather sordid courtroom battle, nominees bought Bentley on behalf of Rolls-Royce Ltd, who also acquired the services of W O Bentley (though not on the design side). The badge reappeared on a Rolls-Royce designed 3.7-liter car, called the 3½-liter (or 'Silent Sports Car') in 1933.

The 3½-liter, in fact, had a new chassis, and individual coachbuilt body styles by coachbuilders, but used a much-modified Rolls-Royce 20/25 engine, and its transmission. Unlike the 'WOs', these were not meant to be racing cars, but gentlemen's carriages – though Eddie Hall raced a 3½-liter in the Tourist Trophy events of 1934 and 1935 with great distinction.

From 1936 the car's engine capacity was increased, from 3669cc to 4257cc, thus creating the 4-liter model. However, this did not do much to raise the 90mph top speeds, or the acceleration, but merely kept pace with the ever increasing weight and bulk of the bodies fitted. In the last two seasons before World War II, the 4¼ became an 'overdrive,' which meant that a

Above: 1929 Bentley 6½-liter appeared too late to save the ailing Bentley company from the financial precipice.

Below: 1939 Bentley 4¼-liter drophead coupe by H J Mulliner – after Rolls Royce had acquired the company.

geared-up top gear ratio was used to allow the cars to cruise, unfussed, on the new highways then being opened all over Europe. All in all, about 2400 of the 3½ and 4¼ 'Derby' Bentleys were built.

Following the war, a new generation of rationalized Bentley and Rolls-Royce cars were produced from the new factory at Crewe, the two types of car becoming increasingly alike. However, one model, the R-Type Continental, announced in 1952, was very definitely a sporting Bentley, for it had special rakish body lines, usually by H J Mulliner, and a top speed of up to 120mph, in great elegance and comfort. Between 1952 and 1955, 208 of these cars were produced, at first with 4¼-liter overhead inlet/side exhaust valves, later with 4566cc units, and finally, from 1954, with 4887cc engines.

The second generation of postwar cars were the Rolls-Royce Silver Clouds and S-Series Bentleys. Continental versions – some with convertible coachwork, some with the 'traditional' two-door coupe style – were always on offer, from the end of 1959 with the all-new light-alloy engine.

Finally, in the early 1980s, came an attempt to revive the 'Bentley' image, when the Bentley Mulsanne saloon was redeveloped, with a massively powerful (unofficially, 300hp) 6750cc V8 engine and marketed as the Mulsanne Turbo. Even at a 1985 price of more than £60,000, this 135mph car caused a real stir, and may be the first of several more individualized Bentleys.

Top right: 1934 Bentley 3½-liter Sedanca by Gurney Nutting.
Right: Fastback Bentley R Type.
Below: 1982 Bentley Mulsanne featured turbocharging rather than the familiar 'blower'.

FGU 30

BMW

The Bayerische Motoren Werke (that means Bavarian Motor Works), or BMW as the marque became known, evolved from a grouping in the early part of this century. One of the companies involved, Dixi, were in such straits that they were wise enough to acquire the licence to assemble British Austin Sevens in Germany. BMW took over Dixi in 1928 and continued to build these cars, rebadging them as BMWs.

The famous BMW badge, incidentally, was a stylized rendering of a propeller at speed, for apart from motorcycles, BMW's principal product was aeroplane engines, many of which were supplied in the 1930s and 1940s to the German air force.

It was not until 1933 that BMW produced their own design of car, the Type 303, which had a 1173cc six-cylinder engine, producing 30hp, and was the first to use the famous 'kidney' style grille, which has featured on every BMW car, in one form or another, ever since. 303 became 315 in 1934 (with 1490cc), and the 1911cc/45hp Type 319 in 1935. It was from this advanced base, which was built on a tubular chassis frame, that more sporting BMWs, culminating in the 328 of 1936, were developed.

The 328 had a streamlined, curvaceous, two-seat body style, independent front suspension, and hydraulic brakes. For this model, too, a new cylinder head was developed, with opposed valves and part-spherical combustion chambers, the valves being operated by a complicated cross-pushrod arrangement. Three downdraught Solex carburetors were fitted, peak power from the enlarged (1971cc) engine was 80hp, and the top speed of every car was nearly 95mph. Well over 100mph was available in competition tune. The cars, however, were always expensive, and only 461 were sold before production ended – due to the war – in 1940.

In its short career, the 328 was a successful competition car. Notably, it won its debut race at the Eifelrennen in 1936, and other examples won the 2-liter class of the 1938 Spa 24 Hours race, and the Le Mans race of the same year. *The Autocar*'s well-known sports editor, S C H 'Sammy' Davis achieved 102.22 miles in one hour, at Brooklands, in 1937 in a 328.

Five very special 328s were built for the Mille Miglia of 1940 (racing went on in Italy until that country joined the fighting), two with streamlined closed coupe bodies, and three with more streamlined open two-seater styles. These cars were further distinguished by having lightweight frames, and their engines had been boosted from 80hp to 135hp. Four of the cars finished, one of them – driven by Huschke von Hanstein (later Porsche's competitions manager) – winning the race outright at an average speed of 100mph.

BMW's postwar car production did not

Left: 1938 BMW 328 with classic lines showing the true sporting heritage of this prestigious marque.
Right: 1940 BMW 327. *Below:* 1937 BMW 319

begin again until 1952, when the 501 saloon was offered, with a prewar type 1971cc engine. The same chassis and bulbous body, however, became the 502 in 1954, with the insertion of a new overhead valve 2580cc V8 engine, which had aluminum cylinder heads and produced 100hp. It was on the basis of this car that BMW later offered two more sporty cars, the 3168cc Type 503 Cabriolets or Coupes, and the short-wheelbase two-seater 507, with a 150hp version of this engine.

The 507, styled by Count Albrecht Goertz, could be supplied as an open car, or as a hardtop coupe. Its chassis was a shortened version of the 501/502/503 design, with mainly tubular and box-section members. Independent front suspension was by longitudinal torsion bars and wishbones, the live rear axle also having torsion bars and radius arm location. In standard form the 507 could sprint to 60mph in about 11 seconds, and in spite of the top speed being up to 135mph overall fuel consumption was estimated at 17mpg. Even so, the 507 was not a commercial success, for it was very expensive. It was dropped in 1959, after only 253 examples had been built.

Having built 603 examples of the bigger and bulkier Bertone-styled 3200 CS in the early 1960s, BMW spent some years building up their new range of saloon cars with four-cylinder overhead camshaft engines. The first sporty car to come from this pedigree was the 2-liter 100 or 120hp 2000C/CS Coupe, which was assembled for BMW by Karmann at Osnabruck. These cars inherited the new BMW six-cylinder engines for 1969, and by 1971 had 2985cc, 200hp, and were known as 3.0CSi.

BMW then got into head-to-head conflict with Ford's Capris, in touring car racing, and evolved the 3.0CSL, which had more power, and many light alloy body panels. The final development of this design was the 'Batmobile,' of which only 39 examples were built, with front and rear spoilers, and other mechanical refinements. In touring car racing these cars were victorious until the late 1970s.

To follow up this success, BMW produced a new series of coupes, the 6-Series models, in 1976. But these were heavier, grown-fat, Grand Tourers. Their thoughts also turned to sportscar racing, so a 'Supercar' was conceived, as an homologation special. Designed, and originally developed for BMW by Lamborghini, the new car, dubbed M1, had a multi-tubular chassis, all-independent suspension, a fiberglass body, and a mid-engine mounting, the unit being a 24 valve 3.5-liter version of BMW's racing 'six,' with 277hp.

Styling had been by Giugiaro, and construction of the cars was originally to have been at Lamborghini, but following that concern's financial traumas, assembly was taken on by Baur of Stuttgart. The M1, though revealed in 1978, was still not ready to go on sale at the time, and to keep up the interest in the meantime, fleets of M1s were used in the Procar series of races, preceding World Championship GP races in 1979 and 1980. This was a very costly publicity exercise involving drivers like Niki Lauda and Nelson Piquet (who won in 1979 and 1980, respectively), but the 500hp racing M1s could reach 200mph, and were very spectacular indeed. There was also a turbocharged example built, which was said to produce 850hp!

The road cars, too, were sensational enough, for they were not only very refined, with excellent roadholding, but could reach 60mph from rest in about 5.5 seconds, and had a 162mph top speed. The last M1 of all – a total of 456 were built – was completed in 1981.

Below: 1958 BMW 507.
Above right and far right: 1983 BMW 3.5-liter Alpina.
Below right: The BMW M1 failed to find its intended mark but gave rise to the Procar Series instead.

AAL 762Y
ALPINA

M-VM 3853

Bugatti

One of the most famous sportscar manufacturers of all was Bugatti. The cars were aesthetically magnificent, if sometimes technically backward, and all were the work of Ettore Bugatti himself. He was born in Italy, and designed his very first car in 1900, when he was still only 19 years old. Later he was responsible for the layout of several other cars not bearing his name, before he started up his own company at Molsheim, near Strasbourg, now in France, in 1909. Even the earliest Bugattis, which had small, relatively hard-working engines, showed signs of the precision manufacture which would always be a hallmark of the owner's quest for excellence.

Even before World War I, the little 1327cc Type 13, and the much larger, chain-driven, 'Garros' model enjoyed success in competition. After the war, Bugatti's first truly sporting cars, the Type 13 'Brescia' models, were evolved. So named because cars of this type took the first four places in the Italian GP supporting race, at Brescia, in 1921, there were various derivatives, with engines between 1368 and 1496cc, with the most powerful of all having overhead camshaft 16-valve four-cylinder engines.

Up to 40hp was quoted, and this was enough to propel the tiny (1350lb) racing versions at up to 100mph. Longer wheelbase touring models (the Types 22 and 23), could maintain 70mph on 10hp fewer. Brescias also won the Le Mans Voiturette race in 1920, and notched up many impressive victories all over Europe. In Britain, for instance, the model became a formidable hill-climb car in Raymond Mays's hands. For the first three years, incidentally, these cars had pear-shaped radiator profiles, but later cars had radiators closer to the classic Bugatti 'horseshoe' style. Made continually until 1926, about 2000 were assembled at Molsheim, though variants were also built by Crossley in England, Rabag in Germany, and Diatto in Italy.

In 1924, Bugatti produced the famous Type 35, that was to achieve some 2000 competition victories in its production life of seven years. There were several sub-derivatives of the Type 35, with engines from 1100cc to 2300cc, and sports or out-and-out racing car specifications. The Type 35, like the earlier Type 30 of 1922, used a straight-eight cylinder engine, originally of 2 liters and 90hp. (No Bugatti, incidentally, ever had a six-cylinder engine.) Later there were 2.3-liter Type 35Ts (T = Targa Florio), and various Grand Prix types, including the supercharged Type 35Bs and 35Cs.

The straight-eight engine was a masterpiece, both in its appearance, and in its function. Two separate four-cylinder aluminum blocks were fixed to a common crankcase, all fits and finish being such that gaskets were not required. The crankshaft itself ran in five ball, or roller, bearings. A Type 35 weighed about 1650lb, and racing versions were dominant in GP racing for several years. Most 'touring' versions were Type 35As, which looked like the GP racing derivatives, but had a less-special three-ball-bearing crankshaft engine from the Type 38. Top speed of these cars was about 90mph, though competition types were capable of well over 100mph – a 35B was timed at almost 125mph in 1930.

The next truly sporting Bugatti road car to be announced (in 1926) was the four-cylin-

Below: 1927 Bugatti Type 35C with Type 35 radiator shell.

Below: 1925 Bugatti Type 35T.
Right: 1925 Bugatti Type 37. In 1925 and 1926 Bugattis scored 1045 different racing event victories.
Below right: 1932 Bugatti Type 55.

MOTORS

GB

2PT

Top of page: 3.3-liter supercharged Bugatti type 57SC, introduced in 1934. About 700 were produced up to 1939, when it was among the fastest stock road cars available.
Left: Bugatti Type 55.
Above: Bugatti Type 57 Atlantique.

der Type 37, which effectively replaced the Brescia, and had a 1496cc power unit of similar detail design to the straight-eight, all in the chassis and bodywork of the familiar Type 35. Purists compared this car unfavorably with the Type 35; the touring version of the design, the Type 40, was rudely described as 'Ettore's Morris Cowley'! Both models were capable of 70mph, and were much less temperamental than the 'eights' – 290 Type 37s and 840 Type 40s were built in a five-year production life.

For 1927 there was a new Type 43, which used the 2.3-liter supercharged eight-cylinder engine from the Type 35B racing car. Often called a 'Grand Sport' model, it could reach 90mph from rest in 30 seconds, and had a top speed of about 105mph. In spite of success in contemporary rallies, it was not a sales success – only 160 were sold in five years.

Type 43, therefore, was replaced by Type 55, in 1932, this being one of Bugatti's most sporting cars of the 1930s. This had a twin overhead camshaft version of the 2.3-liter eight-cylinder engine (actually the Type 51 racing unit), which produced 135hp. Maximum speeds were as high as 112mph. Several different body styles were seen, but only 38 cars of the type were ever built.

In addition to the very sporting Bugattis, there were many high-speed touring cars (many of which were better than the sportscars from rival manufacturers!), one of the most famous of the later types being the splendid 3.3-liter eight-cylinder twin-cam (140hp) Type 57, first seen in 1934. Though inspired by Ettore, this was largely the detail work of his son, Jean, and was sold in various distinctly different types, with many body styles. The most sporting of all were the short-chassis, lowered Type 57S, and its supercharged version, which was the 200hp Type 57SC, though only about 40 of these two derivatives, from a total Type 57 output of 750 cars, were produced. Examples of the Type 57 won the 1936 French GP (a sportscar race), and the Le Mans 24 Hour races of 1937 and 1939.

During World War II, the Bugatti factory was occupied by the Germans, and car production never seriously re-started after that. The only Bugattis sold were a handful of Type 101s, which were effectively Type 57 chassis with updated, rather bizarre, styles.

Below: 1938 Bugatti Type 57S gave 130hp, boosted to 160hp with supercharger.

Chevrolet Camaro

The Chevrolet Camaro, along with the Pontiac Firebird, was General Motor's admitted direct response to the extraordinary success of the Ford Mustang, which had been launched in 1964.

The original Camaro, therefore, was rushed through for announcement in the autumn of 1966, and its unit construction body shell used much of the underpan of the Chevrolet Chevy II, and many of that car's mechanical components – just as the Mustang was closely based on the Ford Falcon. Because they were running later, however, GM wanted their car to be longer, lower, wider and more spacious than the Mustang, and to be faster into the bargain.

In its first year, the Camaro was offered as a four-seater coupe or convertible, with a choice of straight-six, or V8 engines, spanning 140hp and 3.8 liters, to 6.5 liters of V8, with 375hp (gross). Specialist versions, all with masses of options, were the RS, SS350, and Z-28 derivatives. 220,000 cars were built in the first model year, but this was pushed up to nearly a quarter million a year later.

Drum brakes were standard on these cars at first, but disk brakes of Chevrolet Corvette type were made optional from 1967. There were minor styling changes in the next three years, but the really sensational change was held back for the beginning of 1970, when the completely re-styled '1970½' Camaro was launched.

Although the new car still ran on the same 108in wheelbase, it was much more attractive, and had a near-unique body shell instead of having to rely on the Chevy II. It had a more slippery looking shape, was still a 'personal' car, but was afflicted with a cramped rear seat, and a small trunk, which was partly filled up by spare wheel. On this car, too, there was a convertible option – all types having the same coupe style.

The new car's 'base' engine was now a 4.1-liter six, there were four different-tune 'small-block' V8s up to 5.7 liters, topped off by two mighty 6573cc V8s, the most powerful offering 375hp. As in every other year since 1967, there was a high-performance Z-28 model, this being the most effective, if not the most powerful, of all; this was visually distinguished by a full-width rear spoiler, and special paint striping, and had special, firmed-up, suspension. All new-style Camaros were fitted with front-wheel disk brakes as standard, though the rear brakes were still drums.

The new wave of USA legislation – safety and exhaust emission – soon took its toll of the Camaro, so that there were no 'big-block' V8 engines after 1972 (the last of all produced a mere 240hp, nett, while the smaller-engined Z-28 had 255hp).

For 1974, the cars had a striking new 'shovel-nose' body style, which incorporated a streamlined polyurethane soft cover encasing the headlamps, and covering the new style 'Federal' front bumpers; this style was retained until the end of the 1977 model year.

With safety and other legislation in mind, Chevrolet dropped the Z-28 derivative for 1975 and 1976. However, the car still sold satisfactorily, especially after a new wrap-around rear window style was introduced for 1975; this made the cabin much lighter, and more attractive, especially for rear seat passengers.

Sales sagged in the aftermath of the energy crisis of 1973/74, but then began climbing again in 1977, to approach the levels achieved by the original 1960s-type Camaro. Chevrolet never let the car get stale, or overfamiliar; for 1978 they gave it a new nose, and new tail styles, all in body paint colours (this was particularly attractive on the Z-28, which had been re-introduced, where there were added aerodynamic spoilers). In 1979 the prestigious Z-28s were even more striking, with a new set of vents, spoilers and paint jobs.

In 1980, then, this particular Camaro celebrated its tenth birthday (remarkable by American standards, and only beaten at Chevrolet by the current Corvette). In that year, incidentally, the Camaro's 'base' engine became a new 115hp, 3.8-liter V6, while the Z-28 continued to use a 190hp Corvette V8.

This particular body style was finally retired in the summer of 1981, after just less than two million had been built in eleven

Below: The formidable 1967 Chevrolet Camaro SS350 convertible came with a bewildering range of dealer options.
Above: GM's reply to the Form Mustang, the 1967 Camaro with Ram Air 400 option gained a firm place in the musclecar market.

years. It was a better car than ever, when it was retired, and had long since silenced all its critics.

The new Camaro range which took over, however, had been under development for several years, and looked stunning. It was designed not only to look good, but to have a very efficient aerodynamic shape, it had a seven inch shorter wheelbase than the old car, was 500lb lighter, and the bodywork was ten inches shorter.

Although its rear seat was no more spacious than before, the seat squab could be folded down to provide a larger luggage platform, and the new style was a true hatchback, offering an opening rear window. For the first time, here was a Camaro with MacPherson strut independent front suspension, which helped give it outstanding stability and roadholding, and there was a choice of V6 and V8 engines.

The Z-28 version was continued of course, fitted with a fuel-injection V8 engine, and Chevrolet actually produced a Z-28E, where E stood for Europe, took off all the performance-stifling emission gear (including the catalytic converter), and sold it to many other countries, where it excited many testers. The problem was that as the dollar rose in value, so too did the export price of the Z-28E. . . .

For 1985 the Camaro was improved yet again, by the addition of a new fuel injection installation, shared with the Corvette (which had been renewed for 1984). In the IROC Z-28 version, the '85 had a top speed of 140mph, and could leap from rest to 60mph in just seven seconds, all highly anti-social by current USA habits. Nevertheless, in spite of all the latest emission limitations, this was, so far, the fastest Camaro yet made.

Below: Customized 1967 Chevrolet Camaro Z28, introduced with 5-liter powerplant to suit Trans Am rules.

Chevrolet Corvette

If you judge a successful sportscar by the numbers produced, then the Corvette stands head and shoulders above the rest. By 1985, more than three quarters of a million had been made, every one of them with a corrosion-proof fiberglass body shell.

The Corvette was first conceived as a styling 'dream' car, but it was put into production in Flint, Michigan, where 300 were built, using many components from the standard Chevrolet saloon car models. The body style was sleek and futuristic, with faired-in headlamps, and a wrap-around windscreen, but power was by the famous Chevrolet stovebolt six-cylinder engine, tuned to give 150hp.

Other small US manufacturers had experimented with fiberglass bodies, but Chevrolet, as the world's larger car makers, were the first to adopt it for American production cars. GM, seeking to keep costs in order, realized that this would save much, compared with the cost of producing steel dies, and it would also help them to seem advanced, and 'young' minded.

In its first two years the car was not a success, not only because it was too expensive, compared to European imports, but because enthusiasts did not like its two-speed automatic transmission. The 'boulevard' set, too, were horrified to find a car without winding, but only detachable, windows. Even in 1955, when Chevrolet's new 4.3-liter V8 engine (190hp) became available, only 700 cars were sold.

For 1956, however, there was a much revised body style, with proper winding windows, and more purposeful lines. There was the option of a proper hardtop, the most powerful engine type offering 240hp, and – best of all – a three speed manual transmission became standard. The following year, Chevrolet made it even better, with a four-speed manual transmission, and a fuel-injection option on an engine enlarged to 283 cubic inches (4.6-liter), with 283hp!

A face-lifted body style was a 1958 feature, with four headlamps instead of two, much heavier bumpers, extra chrome trim, and an improved instrument panel. For 1959 there were few major changes, though in that year the Briggs Cunningham team achieved eighth place in the Le Mans 24 Hour race, but for 1961 the tail end shape was crisped up, at the same time as a more restrained type of front style was standardized. The '62 was even 'cleaner,' and the engine size went up to 327cu.in/5.35 liters.

All this, though, was a prelude to the striking Corvette Sting Ray, first seen in the autumn of 1962. This car featured a new chassis frame, with all-independent suspension, and a new body style, either as fastback coupe, with divided-pane rear window, or a convertible; the divided pane, incidentally, lasted only one year, becoming a conventional wrap-around for 1964.

For 1965 there were disk brakes all round, and the option of the 'big-block' 6.5-liter/425hp V8 engine. All Corvettes were fast, but some were faster than others. This, incidentally, was the last year in which fuel injection was available, though it was re-introduced many years later, for 1982! The 'big-block' engine, incidentally, was enlarged to a full seven liters (427cu.in) for 1966.

The 1963-model chassis was continued unchanged when the next generation Corvette appeared for the 1968 model year, under a body style that was longer, lower, more streamlined, with pronounced wheel arch bulges. This car had vacuum controlled flip-up headlamps, and a cover for the windscreen wipers, plus a full fiber-optic bulb monitoring system. For 1968 the car was Stingray (all one word).

Until early 1984, this car was gradually changed, refined in some ways, but made much less special in others, yet continued to sell remarkably well. During the early 1970s, of course, not only did the engines

Below: 1956 Chevrolet Corvette convertible showed new styling over the original 1953 car.

have to be de-tuned, to keep up with emission laws but the largest engines were eventually dropped, and (from 1972) lower 'nett' horsepower figures were quoted instead of the more impressive looking 'gross' figures.

To keep abreast with safety legislation, too, the 1973 Sting Ray had a new, longer, nose, which incorporated a soft deformable polyurethane cover over the '5mph impact' bumper. For 1974, similarly-inspired changes were made at the rear, and this was also the last year in which the large-block (now 7.4 liters) engine was available. The last Corvette convertible was built in 1975, killed not only because of a concern over future roll-over legislation, but also because of a lack of demand.

There were two special 'Limited Edition' Corvettes in 1978, one of them being Indianapolis 500 pace car replicas. A significant style change was the introduction of fastback line to the bodywork, which not only improved the looks, but increased the luggage capacity. For 1980 there was a new 'shovel nose,' designed with the help of wind-tunnel testing, along with a new integral tail-end spoiler. High-performance engines were dropped in 1980, while fuel injection became available again in 1982, not only to boost power outputs, but to improve fuel economy.

Finally, after a very long run indeed, the Corvette was renewed in 1983, in a stunning new guise, an all-new chassis and body style. A new factory to build this car was opened at Bowling Green, Kentucky, although the old-model Corvette had moved there, from St Louis, in 1981.

Above: Bill Mitchell's Sting Ray, the 1966 Chevrolet Corvette.

The new model, whose style featured a removable roof panel, and a hatchback rear window, had full digital instrumentation. The body was fiberglass, as always, the whole front section of which could be lifted up to give access to the engine and front suspension. It was a much lighter car than its predecessor, faster than ever, and independent road testers agreed that its cornering power on smooth surfaces was unequalled by any other production car.

There were, however, criticisms of a poor ride and – incredibly for an American car – criticisms of the over-responsive steering. In 1985, therefore, the car was given softer suspension settings, and had better straight line stability. There was also new electronic fuel injection, and an increase in peak power to 230hp, which promised acceleration from 0-60mph in 5.7 seconds, and a top speed of 151mph if high-octane, lead-free fuel was available, but with all emission equipment still installed. The latest Corvette, therefore, had performance which would have been unthinkable just ten years ago.

Cunningham

Millionaire sportsman Briggs Cunningham's greatest ambition was to win the French Le Mans 24 Hours race; his achievements already included victory in the Americas Cup for ocean yachts. For his first attempt on this race, he entered two cars in 1950 – one was a Cadillac two-door coupe with a standard body style (this car finishing tenth to the amazement of the European crowd), the other being his first-ever prototype sportscar, a rebodied Cadillac known as 'Le Monstre,' or the C-1.

For 1951, he returned to Le Mans in strength, with three race cars and a spare of the new C-2R. This car had a massive tubular chassis frame, independent front suspension, de Dion rear suspension with inboard brakes, a 220hp version of the Chrysler Hemi engine, and a three-speed Cadillac transmission. One of these cars was in second place after 16 hours, but mechanical troubles then ensued, and it could only finish eighteenth. The other cars retired.

At the time, Le Mans race regulations stated that prototypes could only be entered by an established motorcar manufacturer so, nothing daunted, Cunningham set up a 'production line,' and produced a series of 18 C-3 coupes, which had the same basic mechanical design, though semi-automatic transmission was also optional, along with coupe bodywork by Vignale.

Cunningham's quest for Le Mans honors was unavailing. In 1952, one of his C-4Rs driven by Cunningham himself finished fourth, while in 1953 the same models achieved seventh and tenth places. In the same year, however, the new rigid-axled C-5 took third place. His last Le Mans attempt was in 1954 when, once again with C-4Rs, his cars achieved third and fifth places. This was remarkable by any standards (don't forget that a Ferrari won, and a Jaguar D-Type came second), but it was still disappointing for a man who always wanted to win outright.

Cunningham continued racing for several years, often using Jaguars painted white with blue striping, but he never again produced his own machines.

Right: 1953 Cunningham 5.4-liter came as a body dropped onto an engine/chassis combination which was often Pontiac, but varied.
Below: Cunningham C2R of 1951 – one of three Le Mans cars.

Daimler SP250

For many years, the Daimler company produced stately, dignified cars, as befitted the marque known for supplying limousines to the British royal family. However, one notable between-wars exception to this was the famous Daimler Double-Six chassis, fitted with low-slung Park Ward body. This car had a 150hp version of the 7-liter sleeve-valve V12 engine, developed by Thomson and Taylor of Brooklands.

After World War II, Daimlers gradually became more personal, and more sporting, their mid-1950s products being known as Conquest, Conquest Century and Sportsman. After a management upheaval, there was yet another change of marketing emphasis. Even so, the launch of a completely new, fast, two-seater sportscar, in the spring of 1959, was a complete break from Daimler traditions.

The engine was designed by Edward Turner, a noted motorcycle designer from the BSA Group; he had taken charge at Daimler two years previously, and his first task was to produce a range of new power units. His motorcycle work was not only respected, but successful, so it was no surprise to see that he incorporated much the same valve gear in a smooth, light, and compact V8 of 2548cc, which developed 140hp at 5800rpm. Compared with any previous Daimler engine, it produced exceptional torque, and was as smooth as any sleeve-unit ever had been.

At first, the new sportscar, shown for the first time at the New York exhibition, was called 'Dart,' but there were immediate objections from Dodge, of Detroit, who already used that name on their own cars. After a threat of legal action, the Daimler name was changed – to SP250. Nevertheless, many enthusiasts still familiarly call this car the 'Dart'!

The SP250 had a conventional cruciform-braced separate chassis frame, which was quite clearly modeled on the Triumph TR3A frame of that period, and it used the same front suspension components. The rigid rear axle was sprung on half-elliptic leaf springs.

Behind the engine was another TR-like component, the four-speed manual transmission, with optional overdrive, or even the option of an automatic transmission. Girling disk brakes were fitted all round, and because the kerb weight was less than a tonne, the car could sprint to 60mph in 10.2 seconds and top 120mph flat out.

The major talking point of the SP250 was its body shell, not only for the unmistakable (some said ugly) styling, but because it was made from fiberglass, the first time this had been attempted by Daimler. At first the shell was too flexible, and there were quality and refinement problems.

The SP250 was designed to appeal to export markets (the USA, in particular), but never sold in the hoped-for quantities, nor was it ever profitable to its makers. Daimler, in fact, were taken over by Jaguar in 1960, who persevered with the SP250, eventually developing two improved versions of it, all with the same styling.

The 'B-Specification' was phased in during 1961, in which previous extras such as the bumpers, the heater, and the adjustable steering column, along with a stiffer body shell, were all included. The 'C-Spec' cars which followed had similar detail changes, but a proposed restyle was canceled, and the last of 2648 SP250s was built in 1964.

Above and below: V8-engined 1962 Daimler SP 250 B featured fiberglass bodywork. Customers included traffic police.

Datsun

The Japanese firm Datsun, which had started by assembling Austin cars under licence, first broke into the USA sportscar market in the mid-1960s with the 1600 Sports, and 2-liter models. Both of these cars were strictly conventional by British and popular American standards. However, they soon began to appear dated, for they had robust but not very refined engines, a ride which was definitely secondary to the roadholding, and a rather stark two-seater style.

The company realized, however, that there was enormous potential in the United States, and developed a new model, the 240Z, specifically for this market. Although the 240Z could not compete on performance with the Jaguar E-Type or Italian exotica, it was a much more modern car than the MGB and the TR6 and soon took over market leadership.

In general layout, except for its independent rear suspension, the 240Z was something of a classic design – some pundits even compared it to a 're-born' Austin-Healey 3000. It used a front engine (2393cc) with rear drive; the engine being an in-line overhead cam six, as already found in other Datsun passenger cars. The style was by a consultant, Count Albrecht Goertz, with modifications by Datsun, had a sleek fastback look, with lift-up tailgate, and was perhaps influenced by the E-Type and Ferrari 275GTB lines. The USA loved it for that, especially as its price was a mere $3700.

In many respects it was a conventional two-seater, for it had a unit construction steel hull, MacPherson strut independent front suspension, allied to rack and pinion steering, and with 10.7in front disk brakes. However, there was independent rear suspension by MacPherson struts and lower wishbones, and the ride was much more supple than that of the displaced Fairlady models.

Some versions were called Nissan, not Datsun, and some had 2-liter instead of 2.4-liter engines, but the vast majority were the affectionately-titled 240Z 'Z-Cars.' They had strong, smooth, responsive engines that could be taken up to a red-lined 7000rpm, and with 150hp they had 122mph top speeds, and a 0-60mph potential of 8.7 seconds. Overall fuel consumption was usually better than 20mpg.

In the 1970s, the European sportscar manufacturers gradually slipped back, or pulled out altogether, principally because they found it too difficult, or costly, to meet the tight new exhaust emission and safety laws in the USA, so this left the market wide open to Datsun, and the Z-Car. By 1972 it had become the world's fastest selling sports coupe.

Its European image took longer to establish, in spite of its good looks, and high performance, but fortunately it was bolstered by a successful competitions program, which included two outright victories in the East African Safari rally, in 1971 and 1973.

Below: 1965 Datsun SP 311; victories in the East African Safari would soon establish the Z-car as one of America's favourites.

Top: 1952 Datsun Sports.
Above: 1960 Datsun SP 212.
Top right: Datsun 260Z.
Bottom right: 280ZX Targa 2+2 came after Datsun adopted the Nissan name worldwide.

The original car stayed in production until 1973, by which time no fewer than 156,076 240Zs had been built. It was then replaced, for 1974, by a derivative known as the 260Z, which was effectively a 240Z with a larger, 2565cc engine, with a longer-wheelbase derivative offering extra occasional seats, logically enough called 260Z 2+2. However, in spite of having more advertised power and torque, the 260Zs were actually slightly slower, and less sporting, than before.

The 260Z remained on sale world-wide for four years, yet in the USA it was displaced by the 280Z derivative after a life of just more than one year. As its title implies, the 280Z had a larger, 2753cc, engine, especially tuned to meet the latest emission laws, and the starting and drivability were all made more acceptable by the use of Bosch L-Jetronic fuel injection. This fitting also helped improve fuel consumption, though it was always a losing battle as the 280Z had put on an alarming amount of weight – it was 500lb heavier than the original 240Z.

The weight penalized performance, which was now down to a top speed of 115mph and 0-60mph in 10.1 seconds, in USA 'Federal' tune. Even so, the 280Z, and its 2+2 derivative, continued to sell well against new opposition (like BL's TR7) – with 70,000 being delivered in the USA in 1977 alone. In the UK, however, it was now behind the times, not only in looks, but in value against cars like Ford's 3-liter Capri.

310,497 of all 260Z and 280Z types were sold before the end of 1978, at which time the range was dropped, in favor of the larger, heavier, 'softer,' and altogether less attractive 280ZX range, which the British buyers did not seem to like at all.

260Z

DATSUN

Delahaye

Today Delahaye is a forgotten marque. Even for those who can remember the name will only recall the fairly fast, well-equipped, French cars which became famous for a time, after the firm merged with Delage in 1935. But Delahaye has a much longer, and more distinguished, history than this, for Emile Delahaye built his first automobile in 1895.

For a while, however, there was a dull period when the company concentrated on marine engines. They produced a mammoth 7.0-liter four-cylinder engine with twin overhead camshaft valve gear in 1903 (but not for cars) – the original motorcar twin-cam, being that seen in the GP Peugeot of 1912.

After a long and not very exciting period, the company completely changed its approach to car design, when a young designer called Jean Francois joined the company in 1934. The result was the altogether more enterprising 135 range, which was in direct competition with the other French marques of Lorraine, Delage and Bugatti.

The 135 had a sturdy six-cylinder engine of 3557cc, with push-rod overhead valve gear, which in sporting tune produced 120hp. For racing there was a special cylinder head, three twin-choke carburetors and peak power output was 155hp. It was an entirely conventional, though very efficient engine, and because a de-tuned version was also used in one of the company's commercial vehicles, it was soon nicknamed a 'lorry engine.'

At first the 135s had Wilson-type preselector transmissions, but this was quickly replaced by a French Cotal transmission, which had epicyclic internals, and electrical gearchange control, by means of electromagnetic clutches. This last was a typically complex French component of the

Left, below and above: 1936 Delahaye 135 six-cylinder. Delahaye were among the earliest pioneers of the internal combustion engine, producing cars from 1900.

day, and found its way into several other cars. A variety of body styles were available, ranging from open sports to elegant Grand Tourers.

The 135 was also successful in competitions, and in 1936 a fleet of them took second, third, fourth and fifth places in the French (sportscar) GP. Just to show versatility, a 135 then went on to win the 1937 Monte Carlo rally. In competition form, incidentally, a 135 cost £995.

Sister car to the 135 was the Type 145, which used the same basic chassis design, but with an overhead valve 4.5-liter V12 engine. This was reputed to be capable of 250hp, a monstrous figure for its day. The 145 was really intended as a two-seater racing sportscar, for long-distance events like Le Mans, or – with all road equipment removed – for use in GP racing against the Mercedes-Benz and Auto-Union cars. On one famous occasion, at the Pau street circuit in 1937, a Delahaye defeated the mighty Mercedes-Benz!

The 165 was a road-going version of the V12 engined 145, with a greatly detuned engine. It appeared at the 1938 Paris *Salon* (motor show), but it is doubtful if more than a handful were produced before war broke out in September 1939.

After World War II, the 135 was reintroduced, and in 1948 it was joined by the 4½-liter Type 175, which had hydraulic brakes and a claimed (though unsubstantiated) 185hp. The 235 of 1951 was the final development of this range, using the now-familiar 3½-liter six-cylinder engine, under modern, all-enveloping, coachwork, now with 152hp and a 100mph top speed. It was, however, the final fling for this marque, for Hotchkiss took over Delahaye in 1954, and Delahaye cars were almost immediately put out of production.

Below: 1952 Delahaye Figoni 235. Ever-increasing taxation doomed Delahaye and the other Grandes Routieres.

Dodge Charger

When the Charger appeared in 1965, it was an immediate success, because the parent company, the Chrysler Corporation, were making a series of fine big V8 engines. These units had powered the Cunningham racing sportscars of the early 1950s, and the 'Hemi' versions (so called because of the layout of the cylinder heads and combustion chambers) could be supercharged, and subjected to brutal nitro-methane fuel, at enormous ratings, for drag racing.

The Charger, therefore, came along as Chrysler's contender for the newly-booming muscle car craze. It was by no means a sportscar by European standards, for it had a four-seater hardtop coupe body style, and weighed 3500lb. The style, however, was dramatic, for there was a full-width rear lamp panel, and concealed headlamps. The biggest Hemi engine was a massively powerful 425hp 426cu.in (7-liter) V8, and fortunately front disk brakes were soon made available as an option.

A completely revised, and much smoother, body style was made available for 1968, while in 1969 there was not only a minor facelift, but even the availability of a down-market 145hp in-line six-cylinder engine.

The eye-catching derivative introduced in 1969, was the dramatically-different Daytona Charger, intended specifically for NASCAR races like the Daytona 500. Using the already-familiar body shell, it also had a sharp, extended, bullet nose.

Below: 1970 Dodge Charger.
Above right: 1969 426 ci Dodge Charger R/T.
Far right, above: 1970 Dodge Charger R/T.

1970

Dodge Challenger

The Challenger was announced in 1969, by its parent company, the Chrysler Corporation, as a direct competitor to the Chevrolet Camaro and – hopefully – the Ford Mustang. Four-seater coupe or convertible styles were available.

As with all such American cars, the 'base' engine was an in-line six-cylinder engine, but the most exciting of the V8s which most people bought was the 426cu.in (7-liter) 'Hemi' unit.

One distinctive detail about the Challenger R/T model was the use of instruments mounted well forward, recessed into the bonnet panel. In 1971, to satisfy the drag racing fanatics, there was also a Challenger T/A, which had extra large rear tires, side exhaust pipes with megaphone ends, fresh air scoops in the bonnet for feeding to the carburetor, quick-release pins holding down the bonnet, special suspension, a rear spoiler, and a blacked out grille and bonnet panel: T/A, in fact, means Trans-Am, after a popular racing series held in the USA at the time.

Because it was last in the field (and because it was also a big, heavy, car), the Challenger was not a sales success. It lost its 'Hemi' engine option in 1971, after only two seasons, and eventually dropped out of production in 1974, when the most powerful V8 engine produced 245hp.

The name, but not the type of car, re-emerged on a front-wheel-drive Dodge in 1981.

Right: 1970 Dodge Challenger R/T convertible.
Below: 1974 Dodge Challenger 440 came right at the close of the musclecar era.

Duesenberg

Fred Duesenberg was born in Germany in 1876, emigrated to North America, and started a business building bicycles. Then, after designing the Mason car in 1906, he and his brother August set up their own racing engine business and factory, supplying special units to Mason in 1912. A year later the Duesenberg Motor Co. was founded to produce all types of engines – racing, marine, and road car varieties. A special 16-cylinder unit powered a Land Speed Record contender up to 158mph in 1919, and in 1921 a Duesenberg won the prestigious French Grand Prix.

The first Duesenberg production car followed in late 1921; called the Model A, it had a straight eight-cylinder 260cu.in (4.25-liter) engine, and was the first-ever North American car to use hydraulic brakes. Less than 500 cars were sold up to 1926, at which point the company was taken over by the colorful entrepreneur Erret Lobban Cord.

Cord then instructed Duesenberg to develop a new, and very exciting car. This car, the Model J, launched in 1928, had a massive straight-eight 420cu.in (6.9-liter) engine, built by Lycoming, which was another company in the Cord group; it had twin chain-driven overhead camshafts, and four valves per cylinder. Its claimed power output, of 265hp, was twice that of any American rival, and was quite enough to propel an open four-seater to a shattering 116mph – unique performance for the period.

In spite of the onset of the Depression, Duesenberg then produced the even more powerful supercharged SJ for 1932, and it was while testing one of these projectiles that Fred Duesenberg himself had an accident, and later died of complications due to his injuries. The SJ itself, of which only 36 were ever built, was said to develop 320hp, could top 130mph, and achieve 0-100mph in a mere 17 seconds.

Even though the J and SJ Models made widespread use of aluminum parts, the regal and luxurious custom-built coach-

Above: 1931 Duesenberg J Victoria with coachwork by Rollston.
Top right: 1932 Duesenberg J Convertible bodied by Murphy.
Below: 1933 Duesenberg SJ Roadster formed the basis of the two SSJ models built for Clark Gable and Gary Cooper.

work invariably fitted gave them an unladen weight of about 5000lb. Customers had to be wealthy to be able to afford one, and many of them – film stars, politicians or industrialists – specified formal bodywork on the 12ft 9.5in wheelbase chassis. The more rare alternative was an 11ft 10.5in wheelbase frame, topped off by sporting coachwork with two seats, and sometimes with an extra 'Dickey' or 'rumble' seat in the speedster tail. Two famous customers for the ultimate Duesenberg, the supercharged SJ engine in a 10ft 5in wheelbase chassis, with sporting coachwork – these were known as SSJs – were the film stars Clark Gable and Gary Cooper confirming it as a true status symbol.

After Fred Duesenberg's death, his brother August became the company's chief engineer, and was responsible for the SJ record car which achieved a remarkable 135.47mph for 24 consecutive hours (and 152.145mph for one hour) at the Bonneville Salt Flats in 1935.

Only about 500 cars in the Model J Duesenberg family were produced between 1929 and 1937, before financial difficulties in the Cord empire brought production to an end.

Over the years, several attempts were made to revive the great name, and the magnificent styling and character of the originals have become ready targets for 'replicar' builders. The latest of these is the Duesenberg II, not a very original title, which is powered by a Lincoln V8 engine, and is made by the Elite Heritage Motors Corporation of Elroy, in Wisconsin. A two-seater 1932-style roadster costs at least $120,000.

Ferrari

Ferrari is one of the most emotive of all motoring names, and has become a household word. The name has become synonymous with 'fast car,' and conjures up a special magic whenever it is mentioned. The aura has been built up, not only by many fast and exotic cars, but by the unmatched postwar success of the blood-red racing cars in almost every branch of motor racing.

Without one man – the autocratic and occasionally eclectic Enzo Ferrari – there would be no Ferrari legend, for he was the superb organizer, and dogged fighter, who had the ability always to surround himself with gifted engineers. Ferrari himself was born near Modena in 1898; his father was an engineer, so it is not surprising that he always had an interest in such matters. Turning down the chance to study engineering at college (later he admitted to regretting this), he worked for several pioneering companies in the Italian industry, before joining Alfa Romeo in 1920. Here he was able to indulge his love of motor racing, becoming a 'works' driver, and finishing second in the Targa Florio in the same year.

From 1923 he concentrated on organization, rather than driving, as his abilities as a tactician and manager were recognized. In 1929, however, he left Alfa Romeo itself, to form his own racing team, the Scuderia Ferrari, and was entrusted with the task of running the official Alfa Romeo team cars. This arrangement worked well, for it allowed the parent factory to concentrate on designing and developing their next generation of cars. It was at this stage, too, that the famous 'Prancing Horse' emblem first appeared on the cars, and it has been used ever since.

All went well for the team until the mid-1930s, when the German government-financed Mercedes-Benz and Auto-Union GP cars arrived, after which the Alfas could rarely match them again.

After a five-year break in activity, because of World War II, Ferrari's links with Alfa Romeo were severed, and he resolved to start building his own cars, for the first time. His first chief engine designer was Gioacchino Colombo, who produced the classic V12 engine which was to be the mainstay of Ferrari road cars, in so many different forms, for a great many years to come. The original V12 engine was a 1500cc single overhead cam design, first seen in 1947, and cars using it were known as 125s. The first such Ferraris were a two-seater sportscar, and a new Formula One car.

All Ferrari road cars are titled in numbers and letters. The 'system' has become very confusing, and not always logical, over the years. For the first cars, however, the vehicle type number approximated to the capacity of one engine cylinder in cubic centimeters. (However, from 1957, the system used instead indicated the engine size, in liters, and the number of cylinders. Except when it didn't. . . .)

Although Colombo's influence persisted at Ferrari into the 1960s, he left the company in 1950, when he was succeeded by Aurelio Lampredi. In the meantime the V12 engine was enlarged, to 1955cc, and fitted into a car with a very stark style of sportscar body. Known as the Tipo 166, its cycle type wings could be removed, along with the headlamps, and it could be used for Formula racing. It was quick, and competitive, notching up many successes, especially in long-distance events.

After Biondetti's car won the 1948 Mille Miglia, the model was given the suffix 'MM' to its title. This car had a three-Weber carb and a 150hp-at-7000rpm-version of the 60-degree V12 power unit. Its chassis was made from oval-section steel tubes, and featured double wishbone independent front suspension, with a transverse leaf spring, while the rigid rear axle was sprung on half-elliptical leaf springs, with radius arms for location. There was no synchromesh in the transmission, but there were hydraulic brakes. All-enveloping bodywork was produced for the 166 by Carrozzeria Touring, and the car was capable of 125mph. A later version of this car, with

Left: 3-liter V12 Ferrari 250 GT Berlinetta; Pininfarina coachwork established sportscar lines for decades to come.
Above: 1967 Ferrari 275 GTB/4 was marketed in 1966 and raced by works and private entrants.
Right: 1946 Ferrari 166 Corsa Spyder.

an enlarged engine, was the 195 Inter, and this was soon succeeded by the 2562cc 212 Inter of 1951, with a 170hp engine. Shorter-wheelbase types became known as the Sport, or Export, models.

In 1951, however, the first of the really big-engined Ferraris, the 4101cc Type 340 America, was put on sale. This used a new type of V12, familiarly known as the Lampredi (as against the Colombo) type, which produced 220hp at 6000rpm, and gave the

Below left: 1947 Ferrari Type 166 Inter.
Below right: 1950 Ferrari 166MM adopted the suffix after Biondetti won the 1948 Mille Miglia.

Above: 2.4-liter V6-engined Ferrari Dino 246 GT Spyder was introduced in 1969. This one dates from 1973.

car a top speed of 137mph. Further variants, known as the 340 Mexico, and the 340MM, were raced with great success in the United States.

In 1953 and 1954, the factory raced even larger versions of this car, known as the 375 (4522cc) and later still they used 4954cc engines. In the last form, the engine produced 344hp at 6500rpm giving the 375 a top speed in excess of 160mph, with 0-60mph acceleration in just seven seconds.

The first of the famous 250s, the 250 Europa, appeared at the Paris Show of 1953, with a 'long' Lampredi engine, in a 9ft 2in wheelbase, still with the same basic suspension layout. The 250GT Europa which followed in 1954, however, had a shorter (8ft 6in) wheelbase, and a 3-liter version of the Colombo V12. About 50 cars of the two types, in total, were produced.

Then, in 1956, came the 250GT Coupe, mainly with bodywork by Pininfarina, and about 500 of these cars were constructed, for Ferrari was expanding his road car facilities. The 2953cc V12 engine was much as before, but produced 240hp at 7000rpm, using three Weber carburetors. Later models were fitted with Dunlop disk brakes.

One of the all-time classic Ferrari sportscars now followed – the short-wheelbase 250GT Berlinetta, with Pininfarina style coachwork built for Ferrari at Scaglietti, from 1959 to 1962. It was really a racing sportscar, had a 280hp V12 engine, four-speed all-synchromesh transmission, disk brakes all round, but much the same basic chassis and suspension as other Ferrari road cars. In the UK it was made famous by Stirling Moss, who had two successive Berlinetta victories in the Tourist Trophy races.

Two other 250 theme variations of note were the open model known as the Spyder California (announced in 1958), and the 250GT 2+2 (also known as the GTE), the first-ever Ferrari to offer even occasional rear passenger seats. However, the sleekest and most sophisticated Ferrari, up to that time, was the beautiful Pininfarina-styled 250GT Berlinetta Lusso – the word Lusso meaning 'Luxury.' Although the mechanical specification was very similar to that of the SWB Berlinetta, the engine was de-tuned to a mere 250hp.

During the 1960s, the 'basic' Ferrari power unit grew to four liters, and then to 4.4 liters, the first of this new breed being the 330s of 1964. The engine layout, of course, was the inevitable V12, now producing 300hp at 6600rpm, though the chassis was still like that of the 250, with coil spring and wishbone independent front suspension, and a rigid rear axle on leaf springs at the rear, along with four-wheel disk brakes.

The original 330GT 2+2 had four headlamps, but was superseded a year later by the Mk 2 which reverted to two lamps. There was also a differently styled two-seater 330GTC from 1965, and a GTS open Spyder a year later. All were styled for Ferrari by Pininfarina, but built by Scaglietti.

In the meantime, a number of very large-engined Ferraris had been produced, in very limited numbers, starting with the 4.9-liter Tipo 410 Super America of 1956, and the 3967cc Tipo 400 Super America of 1959. Then, to top it all, came the amazing Tipo 500 Superfast, of which only 28 LHD and eight RHD examples were made from 1964 to 1966. It has a 4961cc V12 engine, rated at 400hp at 6500rpm, and was a monstrously powerful and self-indulgent car, the type of machine of which Ferrari legends are created!

A new generation of Ferraris was ushered in with the 275GTB (Coupe) and 275GTS (Spyder) models, of 1964. An all-new multi-tube chassis frame not only incorporated a five-speed transmission, in unit with the final drive, but there was also independent suspension for front and rear wheels, by coil springs and wishbones in each case. The bodies, as ever, were Scaglietti-produced to Pininfarina designs, and the familiar V12 engine had been enlarged to 3286cc, rated at 280hp for the Coupe, yet 'only' 260hp for the open car.

Above: 1972 Ferrari 365 GTB4 Daytona.
Below left: 1975 Ferrari Dino 308 GT/4 2+2. The 'Dino' is named in memory of Enzo Ferrari's son.
Below right: Cockpit, 1976 Ferrari 308 GT4 2+2.

After only two years, this car was updated, to 275GTB/4 specification, the '4' denoting the use of four-cam, rather than two-cam, V12 engines, with power now up to 300hp at 8000rpm, with six twin-choke Weber carburetors.

During the 1950s, Ferrari had also produced a V6 configuration engine, named 'Dino' in memory of Ferrari's own son, who died in his 20s. This unit, in several forms, found many uses, in Formula Two, Formula One, and sportscar racing. Eventually, with the collaboration of Fiat, who took over the 'productionizing' and actual assembly of engines, it was made available in 1986 form for the new mid-engined Dino 206GT of 1967 (as well as for Fiat's own front-engined Dino).

In the 206GT, which had a conventional multi-tube frame, with all-independent coil spring suspension, four-wheel disk brakes and rack and pinion steering, the Dino engine was mounted transversely, behind the two seats, the whole five-speed transmission unit and final drive being driven by spur gears from the crankshaft, and bolted behind the sump. In this form the four-camshaft engine produced 180hp, and the sleekly styled coupe, with many aluminum panels, could achieve 140mph.

The 206, however, had a short life, being succeeded by the 246GT in 1969, in which the wheelbase was slightly increased, the

Left: Ferrari 512i BB replaced the 365 GT/BB in 1976. This is a 1982 model.
Above: 1984 Ferrari Mondial Cabriolet.

engine block changed from aluminum to cast iron, the size increased to 2418cc, and peak power rose to 195hp at 7600rpm. This model continued successfully until 1973/74, and was joined by the 246GTS, almost identically styled, but with a removable roof panel, in 1972.

By this time, however, what many people believe to be the ultimate front-engined Ferrari, the 365GTB/4 Daytona, had appeared in 1968. Using the same basic chassis layout as the superseded 275GTB/4, the Daytona used a 4390cc four-camshaft V12 engine, fed by six Weber carburetors, which produced no less than 352hp at 7500rpm. The transmission was at the rear, in unit with the final drive, the body construction was by Scaglietti, and the Daytona had a top speed of no less than 174mph.

There was also the very rare open Spyder version, known as the 365GTS/4, and in recent years several coupes have been converted to open Spyder specification. One way to pick the age of a Daytona, incidentally, is that from mid-1971 all cars were fitted with retractable headlamps, whereas original European-spec. examples had lamps positioned behind a big clear plastic cover.

The direct replacement for the 246GT series, announced in the autumn of 1973, was the 308 range, starting with the GT/4 2+2, which broke new ground by being the very first Ferrari production car to be styled by Bertone. This wedge-nosed creation had little of the grace of its predecessors, but did, at least, offer +2 seats. Compared with the earlier car, too, there was a longer wheelbase, and a brand new 90-degree 2927cc V8 power unit.

Left and below: 1981 Ferrari Mondial 8 was followed by the 'new' Testarossa and 'new' GTO.

Matters, however, were improved in 1975, when the Pininfarina-styled 308GTB, a true replacement for the 246GT, arrived. Much influenced by the lines of the new Boxer, described next, it had the same chassis and wheelbase as the original car, and for the first two years had fiberglass bodywork. Not surprisingly, a 308GTS (Spyder) soon became available. All the 308s, in their original form, had a top speed of more than 150mph.

The greatest of all Ferrari Supercars, not even discounting the Daytona, was probably the Boxer, which first went on sale in 1973. Correctly entitled the 365GT4 BB, the car had a multi-tube chassis, with all-independent suspension and disk brakes, as one might expect, but it also had a mid-mounted horizontally-opposed 4.4-liter 12-cylinder engine, fed by four Weber carbs, and producing no less than 360hp at 7500rpm. The car's claimed maximum speed was 180mph-plus, but no independent tester ever achieved more than 162mph. The style, by Pininfarina, had swoopy wing lines, and was very like that of the 308GTB which soon followed it.

From mid-1976 the design was updated, to become the BB512. This car had a larger engine, of 4942cc, and a claimed maximum of 188mph, though peak power was down to 340hp at 6800rpm. It was still a phenomenal car, though the claims for its pace were grossly exaggerated, and it sold steadily (more than 150 cars a year) until the mid-1980s.

In 1981, the 308GT4 by Bertone, was replaced by the Pininfarina-styled Mondial 8. Mechanically there were few changes, but the new style was more in keeping with Ferrari traditions, and proved to be a success. Further changes followed in the early 1980s, first with fuel injection being fitted to the engines, then in 1983 all the 308s inheriting four-valve per cylinder heads. The latter cars became known as Quattrovalvole (QV), and the peak power output was 240hp.

Any car which had to replace the Boxer had to be very special, and the Testarossa, launched in the autumn of 1984, was certainly that. The basic mechanical layout was that of the BB512, but the engine had four-valve heads, and power was boosted to no less than 390hp. The Pininfarina style was certainly controversial, for there were slats in the body sides, channeling air into the radiator air intakes behind the doors, and as before, only the one, two-seater Coupe, arrangement was available. Its price, when put on sale in the UK for 1985, was no less than £62,666.

Fiat

Italy's largest and most famous volume producer of cars, has not made many sportscars over the years, but all have been particularly interesting designs. The Societa Anonima Fabrica Italiana di Automobili Torino – a name fortunately shortened to FIAT – was formed in 1899. Like many such pioneering concerns, Fiat chose to publicize their machines in motor sport. They were very successful in 1907, not only when Felice Nazzaro won the Targa Florio, but the Kaiserpreis and the French GP.

The first Fiat roadgoing sportscar, however, was the Balilla, a car derived from the small saloon design of the same name. Balilla, incidentally, was the name given to Mussolini's young soldiers. The Balilla Sports had a four-cylinder, overhead valve 995cc engine which produced 36hp. Because it only weighed 1350lb, and had a four-speed transmission, it offered lively performance. There were a number of modern refinements, including hydraulic brakes. First shown in Milan in 1933, at a price of 14,900 lire, it caused a sensation among buyers. Later, a Mille Miglia model was added to the line-up, this having 'torpedo' body lines, and a more powerful engine.

After the short-lived 1100S and ES (born out of the very specialized 508CMM coupes of 1937) had been introduced in 1947, the first true post-war sporting Fiat was the 8V, shown for the first time at Geneva in March 1952. Power was by a 70-degree 2-liter V8 pushrod engine (that is a very odd vee angle by any standard), either as 105hp or 115 hp tunes, and top speed was 120mph.

The 8V had a tubular chassis, and there was front and rear independent suspension by coil springs and wishbones. Fiat built most of the bodies themselves, as wind-cheating fastbacks, usually with staggered seating, as the cockpit was quite narrow, though several specially styled coachbuilt bodies were also sold.

Only 114 cars were built in all, in two years, and the 8V's most notable achievement was to win the Italian sportscar championship in 1956.

To follow the 8V, Fiat then built a series of cars based on the basic underpan and engineering of the ubiquitous 1100s, 1200s and 1500s, all having coil spring front suspension, and rigid rear axles. The Transformabile of 1955-59 was not a success, becase of its strange styling, but the cabriolets built between 1959 and 1966 sold in large numbers. Fiat styled and built all the bodies, and most had pushrod engines of one Fiat type or another. However, the truly exciting derivative of this design was the 'Osca'-engined device, a 1491cc/80hp unit at first, but a 1568cc/90hp engine from late 1962. The engine was a purpose-built twin-cam, by Osca, though built by Fiat, which paved the way for later Fiat twin-cams. A five-speed all-synchromesh transmission was standardized from March 1965.

Although one must not forget the smart coupe and Spyder versions of the rear-engined 850 models, built from 1965 to 1973, which had a 90mph top speed, the cars which sold most strongly were the two types derived from the 124 saloon, the Sport Spyder and the Sport Coupe, both revealed in the winter of 1966/67. The coupe was on a standard saloon underpan, with Fiat styling, and four-seater accommodation, the Spyder had an open two-seater sportscar shell styled and built by Pininfarina. Both cars, when launched, had 1438cc twin-cam

Above: 1955 Fiat Pininfarina-bodied V8 coupe.
Right: 1935 Fiat 508S Balilla Sport. The Balilla Coppa D'Oro was, along with a small 1100cc saloon, what kept Fiat strong in the years immediately before the outbreak of war.
Below: Fiat Dino Spyder.

engines, and some (not all) had the five-speed transmission first seen in the previous 1500 and 1600 cabriolets.

Over the years, the engines were enlarged, first to 1608cc, then to 1756cc (and alternatively to 1592cc), five-speed transmissions were standardized, along with minor styling changes, but the Spyder carried on alone after 1975, eventually received a 1995cc version of the twin-cam engine, sprouted a turbocharged derivative for USA sale, and was finally hived off to be relaunched as a 'Pininfarina Spyder Europa' in 1982, and is still made in that guise. By most sportscar standards, sales have been enormous.

Between 1972 and 1975, there was a lightened, specialized, version of the Spyder, by Abarth, with independent rear suspension and (for the last years only) with the option of four-valve cylinder heads, which was used almost exclusively for rallying.

To help Ferrari achieve an 'homologated' engine for use in Formula 2, Fiat then took over the V6 Dino engine, redesigned and productionized it, supplied engines to Ferrari, and also launched their own front-engined cars known as Fiat Dinos. There were Spyders, by Pininfarina, and longer-wheelbase coupes by Bertone (final assembly being at Fiat), both having 2-liter four-cam engines, five-speed transmissions, but rigid rear axles. In spite of the 127mph top speeds, this car was only partially successful, no doubt because of the high price.

Nearly 5000 Dinos, mostly coupes, had been built.

Three years later, the Mk 2 Dinos were put on sale, this time with the 2.4-liter iron-block version of the engine, a new ZF five-speed transmission, and coil spring semi-trailing link independent rear suspension (from the 130 saloon). Assembly of the 2.4s was carried out at the Ferrari factory at Maranello, for Fiat had taken a 50 percent stake in that business in 1969. 2398 coupes but only 420 Spyders, were made before production finally ran out in 1973. The engine, incidentally, was also to be used in the Lancia Stratos rally car of 1974-75.

To replace the long running Fiat 850 Spyder, the company next took up a mid-engined project study, proposed at first as a private venture by Bertone, and this appeared as the X1/9 of 1972. Bertone and Fiat, working together, used great ingenuity to package the little two-seater well, so that there were luggage compartments at front and rear. The wedge-nosed style was quite unmistakeable, and the transverse engine/transmission was in the mid-position, behind the seats, having been 'lifted,' almost unaltered, from the front-wheel-drive saloon of the period.

The first X1/9s had 1290cc engines, but from the end of 1978 this was increased to 1498cc, and linked to a five-speed transmission. The latter had 85hp, and gave top speeds of 105-110mph. Clever design features included the use of all-round independent suspension, a removable roof panel, which could be stored in the front luggage compartment, and a spare wheel housing in a compartment behind the seats, accessible through the passenger compartment. Like the 124 Spyder, this was hived off in 1982, and later became a 'Bertone' instead of a Fiat.

Left: 1969 Fiat 124 Sport Spyder.
Right and below left: Mid-engined Fiat X1/9 Targa by Bertone with first 1300 and later 1500cc engine.
Bottom right: 1971 Fiat 124 Sport.

Ford GT40

In the early 1960s, Ford of Detroit committed themselves to an all-out campaign to achieve supremacy in motor sport, and as a beginning tried to take over the Ferrari business. After this foundered, in 1963, they started a sports racing project of their own, to be developed alongside those to win Indianapolis with Ford-powered Lotus single-seaters, and to win the Monte Carlo rally with special Falcons.

The racing sportscar program was located in the UK at first, where Ford hired Eric Broadley of Lola, and bought his 1963 Ford V8 powered prototypes. The car was to be designed at Slough, but the body style, and the V8 engines, were to be supplied from Detroit.

The budget was effectively limitless, and although Ford also attracted John Wyer, from Aston Martin, to manage the effort, it took time to achieve success. Len Bailey was closely involved with the design of the structure, but although the cars were very fast in a straight line, there were aerodynamic, and transmission problems. The first Le Mans race entry was a fiasco.

For 1965, much of the race and development effort was transferred to Shelby American (in California) and Ford's own Kar Kraft subsidiary (in Detroit). But even though they quickly developed the Mk II GTs, completed with 7-liter V8s to replace the original 4.7-liter models, the second Le Mans effort also failed.

Everything began to come right in 1966, however, when the Mk IIs finished first and second at Le Mans (they were actually attempting a dead heat!). In the meantime, a series of GT40 production cars had been built at Slough, and homologated, and it was from this model that the Mk III 'road car' was produced, in which there were adjustable seats, softer suspension, a mild (300hp!) engine tune, and a central rather than right-hand gearchange. Very few of these cars, which can be identified by their four circular headlamps (instead of two rectangular lamps) were ever made.

In the meantime, Kar Kraft had developed the 'J-Car,' which used an aluminum rather than a steel chassis monocoque, and it was this car, with a more aerodynamic type of coupe body, which became the Mk IV, to win the Le Mans race in 1967.

At this point Ford of Detroit decided that they had made their point, and pulled out of sportscar racing, leaving John Wyer's own JW Automotive concern, backed by Gulf, to carry on with the homologated GT40s in 1968 and 1969. Not only did these cars win Le Mans twice more – 1968 and 1969 – but they also won the World Sports Car championship in 1968, before John Wyer also gave up the Fords, and moved over to manage a Porsche 917 team with quite outstanding success.

More than 100 GT40 'production cars' were produced by Ford Advanced Vehicles at Slough, of which 31 were 'road cars.' All the survivors are now valuable collectors' pieces. In recent years, more than one manufacturer has set out to produce convincing replicas of this famous machine.

Right: Ford GT40 road car.
Below: Ford GT40 at Le Mans.

Ford Mustang

Following impressive showings at Indianapolis and in the Monte Carlo rally, Ford's 'Total performance' campaign reached a peak with the announcement of the Ford GT race program. To follow up, the sporty Ford Mustang was launched to reap sales on the back of this glory.

Fortunately for the product planners, who could still recall the Edsel disaster of a few years earlier, the Mustang project was triumphantly right, with the first million sales being notched up in two years.

The Mustang had a simple unit-construction body/chassis unit, with just two doors, and only limited space for rear seat passengers. The style included a very long bonnet, but a stubby boot lid, a high-mounted front grille, and the galloping horse motif that was to become famous.

Much of the chassis engineering was lifted from the Falcon saloon, and there was a range of engines from the 2.8-liter straight six-cylinder, to the 4.7-liter/271hp V8, in the first model year, together with a wide choice of transmissions, extra instrumentation, and front wheel disk brakes.

At first there was a choice of hardtop coupe and convertible bodies, and a fastback hardtop version followed before the end of 1964. The last style was used by Carroll Shelby as the basis for his much-modified GT350 Mustangs – whose specification included 350hp engines, uprated suspension, a limited slip differential, and a price tag of a mere $4547. Ford also offered a further tune-up, the 'Cobra' kit, which included four Weber carburetors and a peak of 343hp. The competition version also included a 34 gallon fuel tank.

The second Mustang generation arrived in late 1966, when the body was reskinned, given a hooded grille, and somewhat more rounded lines. Engines were more powerful than before, and the Shelby monster was now the GT500, with the 428cu.in/7-liter V8 engine, this also being optional on the standard cars, where it gave a 0-60mph time of 5.9 seconds.

For 1969 there was a major restyle, with a longer, wider and altogether heavier and bulkier body. Four headlamps instead of two was one recognition point, and for the first time there was a 'Mach 1' version of the high-performance fastback versions, available with a choice of engines, and an air-cleaner protruding through a hole in the bonnet top. In 1970 a 5-liter/302cu.in engine was made available, so that the Mustang could be eligible for the Trans Am Championship. Need we say, the car was dominant in that series. . . .

The Mustang was re-styled, and enlarged yet again for 1971, even larger, smoother, and more expensive than before, but sales were way down as the US 'pony car' boom had passed on, and the cars were faced with much competition from GM and Chrysler.

It was the last of the truly sporting, and distinctive, Mustangs, for in the autumn of 1973 it gave way to the Mustang II, which was a much smaller, and slower, design, having none of the previous model's appeal, and fitted with the option of four-cylinder or 2.8-liter V6 engines. Nevertheless, it was right for the market, which was gradually turning toward economy, and sold very well indeed, as did the much more sharply-styled Mustang III announced for 1979. By this time, incidentally, a 5-liter V8 engine had once again become optional, though its peak power was down to 140hp, a mere shadow of the original car's capabilities.

Below: Customized version of a 1970 Ford Mustang, Ford's success story from the mid-sixties.

Ford Thunderbird

The first Thunderbird was launched in 1954, as a 1955 model, aimed at the buyers who might be, but were not, buying the controversial Chevrolet Corvette which had been on sale for two seasons. Compared with the Corvette, the Thunderbird was much more attractive. It had an attractive, if conventional two-seater steel body shell with winding windows and an optional hardtop, an an overhead valve V8 engine.

It also had the benefit of Ford's long-established reputation for providing high performance from their V8s, and there was the option of manual transmission as well as an automatic. A rev counter was standard, and the V8's capacity was 256cu.in/4.2-liter.

All the 1956 model year T-Birds had a 'continental' spare wheel kit (which meant that the spare was vertically mounted behind the rear bodywork), while the famous 'portholes' were an option in the hardtop. There were more engine options, and most powerful being the 225hp V8.

There were tail fins for 1957, but this was the last year for the two-seater Thunderbirds, and for 1958 a completely different, longer, heavier, four-seater car was offered, its only sport being provided by the most powerful 300hp engine.

This type of Thunderbird was 'personal' rather than 'sports,' but fortunately it was restyled for 1961, with much more simple lines, a long sweeping wing crown line, and a high front bumper. There was still a convertible option, and for 1962 and 1963 it was offered with a plastic tonneau cover to enclose the rear seats.

The restyling for 1964 was successful, though more decorated than before, and this ran through until late 1966 when a new range appeared, complete with a four-door Landau saloon having 'suicide' rear doors, which were rear-hinged, and with fake cabriolet irons on the outside of the fixed roof. This finally put paid to any pretensions that the Thunderbird has as a sporting car, and the next 15 years of production deserves no mention here. For 1983, however, there was an entirely new type of Thunderbird, having much in common with the latest Lincoln Continental 'Mark' model. This featured a smooth, low-drag, aerodynamically styled four-seater two-door coupe body, with turbocharged 3.8-liter V6 engine, power assisted rack and pinion steering, and variable-height pneumatic suspension units, which bid fair to restore much of the glory to the Thunderbird name.

Below and bottom: 1956 Ford Thunderbird with full Continental kit. A year later the Thunderbird became a four-seat 'personal car'.

Frazer Nash

Before and after World War I, there was a vogue for a cheap and cheerful type of transportation, known as the cycle car. Archie Frazer Nash, and his partner Ron Godfrey capitalized on this by producing a machine known as the GN, which had a twin-cylinder engine in a very rudimentary, basic, chassis frame. It differed from other types, however, in having chain drive rather than shaft drive.

This car was a great success for a time (there was even some manufacture in France). It offered extremely good performance but only basic weather protection. This, and many other cycle cars, met its match in 1922 when the Austin Seven was announced, for that covered the occupants properly, gave them a four-cylinder engine, and most of the 'big car' comforts. At this point Frazer Nash and Godfrey went their separate ways, with Frazer Nash deciding to build his own cars, at Kingston upon Thames, from 1924.

The first Frazer Nash caused a sensation, for it was an out-and-out sportscar, initially fitted with a Plus Power four-cylinder engine (though shortly a side-valve Anzani would be fitted instead). It had stunningly attractive aluminum body styling and – like the GN – it had chain drive transmission. Every Frazer Nash built up to 1939 was to retain chain drive, by which time it was unique, and considered very odd. However, it had the supreme advantages of being very light and offered amazingly rapid gear selection. Early cars were known as fast Tourers, or Super Sports, and 149 were produced before the company ran into financial difficulties in 1929.

At this stage Frazer Nash was rescued by the Aldington family, who had already been involved in selling the cars through their own business, Aldington Motors. When H J Aldington gained control, he was quickly joined by his two brothers, set about revitalizing the company in new premises at Isleworth, and created the 'New Frazer Nash.' This was notable for the use of the overhead valve 1500cc Meadows engine (though the supercharged Anzani was retained for competition cars), and had a painted aluminum, or fabric covered steel body called the Sportop.

The Aldingtons, above all, were very good salesmen, and were always ready to change the car's model name, often with only a few modifications over the previous car. They also realized that it was good policy to name a model after a recent competition success, which explains why names like the Boulogne, Colmore, Exeter, and Nurburg all appeared in sales leaflets.

The best-known model of the 1930s, however, was the TT Replica, first introduced in 1932, but available until 1939. This really looked the part, as a stark, muscly,

sportscar. with its fully louvered bonnet, stone guard over the radiator, a full range of instruments, and the gear lever and hand brake lever both outside the bodywork. It had a characteristic 'bath-tub' style of rear body, with a protruding petrol filler. When observed, with the narrow rear track also on view, the car had a most individual appearance.

Three different engines were offered in the TT Replica – the four-cylinder Meadows, the six-cylinder Blackburne, and a revised and supercharged version of the Meadows. This last was usually called the 'Frazer Nash' engine, though it later became known as the 'Gough' after the man who designed it. A normal Meadows engine developed 55hp at 4500rpm, which gave the car a maximum speed in excess of 80mph.

The cars were very successful in sport, and in particular they shone in the Alpine Trial. In 1932 two cars were entered, and

Left: 1926 Frazer Nash Super Sports.
Bottom left: 1929 Frazer Nash Super Sports.
Right: 1926 Frazer Nash Fast Tourer.
Below: Aluminum-bodied Frazer Nash 'Chain Gang'.

PK 4481

both finished without losing any marks. In 1933 no fewer than seven cars entered, one of which (driven by Aldington and Berry) being the only car in the entire entry to lose no marks at all. In 1934, of the six cars entered, four finished without loss, one lost a single mark, and the sixth, driven by the Hon. Peter Mitchell Thompson, lost 14 due to an engine water pump problem. As a result of this, the team prize was lost to BMW.

H J Aldington became very impressed by BMWs, and lost no time in becoming the British concessionaire for that marque. The result was that by 1935 these cars, known as Frazer-Nash-BMWs, were being sold alongside the chain driven cars, and very little further development went into the domestic models. All in all, a total of 174 'chain-gang' cars were produced at Isleworth, the peak period being in 1932 and 1933, when 32 cars were built in each year.

After World War II, the Aldingtons were instrumental in bringing the BMW designs over to the United Kingdom, where they were revised, and put to very good use by Bristol. Frazer Nash might once have become a part of the Bristol Cars business, but escaped, and announced the first postwar Frazer Nash in 1948.

This was completely different from the 1930s variety, for all models were fitted with the new six-cylinder Bristol (was BMW) 2-liter engine. Once again the practice of naming models after competition successes was continued, thus the postwar range included the Mille Miglia, Targa Florio, Sebring and Le Mans Replica models.

The latter was much the best known, for 34 were built out of a total postwar total of 84 cars. This car was a logical development of the BMW 328 type, but it harked back to the chain-driven cars by being fitted with a narrow two-seater body shell, with exposed wheels and cycle-type wings, and extremely basic weather protection. It was widely (and successfully) used in competitions by such well-known drivers as Stirling Moss, Roy Salvadori, and Tony Crook.

In 1951 a Le Mans Replica won the Targa Florio outright, and in 1952 another example won the very first Sebring (USA) 12-hour race. A Mark 2 version was produced in 1952, and Tony Crook used one to take third place in the 'round the houses' 2-liter race at Monaco. In standard form the Bristol engine produced 110hp at 5250rpm, and the car's top speed was 120mph. More power was available with race tuning, and the superlative roadholding often allowed it to compete successfully against much larger-engined cars.

During the early 1950s, Frazer Nash were appointed UK concessionaires for Porsche cars, and this soon led to the running down of Frazer Nash assembly. The last Frazer Nash of all was produced in 1957, and fitted with a V8 BMW engine, but it was never put into production. AFN, the concern which built Frazer Nash cars after the war, was finally taken over by Porsche in the 1970s.

Left: 1929 Frazer Nash Super Sports.
Top right: 1950 Frazer Nash Le Mans which provided most of the company's postwar racing success.
Below right: 1950 Frazer Nash Le Mans Replica.
Below: 1932 Frazer Nash Exeter.

Hispano Suiza

The unusual name of this famous marque was inspired by two countries – Spain, where it entered production in 1904, and Switzerland where its designer Marc Birkigt was born.

At first there was nothing to make a Hispano-Suiza stand out from dozens of other cars being made at this time, and production in Spain was very small. However, by 1907 several of the more luxurious types had found their way into the garage of the King of Spain, Alfonso XIII.

Following the success of a racing voiturette type of Hispano-Suiza in the French *Coupe de L'Auto* race of 1910, it was decided to name a sports touring version of the car after the King. The 'Alfonso' Hispano-Suiza which evolved over the next two years was a beautifully proportioned machine, usually fitted with a 3620cc engine which developed 64hp at 2300rpm. Because of its side-valve 'T-head' arrangement, and the very long stroke (80mm bore, 180mm stroke) configuration, it had tremendous low-speed slogging power. At first a three-speed transmission was standard, but this was later replaced by a four-speed transmission. At least 75mph was available.

In 1911 the company moved to expand, out of Spain, opening a factory at Levallois-Perret, close to the lucrative markets of Paris, in France. Following the outbreak of World War I, an aero-engine factory was established at Bois-Colombes, and after the war this became the home of the most exotic Hispano-Suiza 'vintage' cars.

The Spanish factory, however, continued to make cars, until after World War II some of its premises and personnel became involved in the Pegaso project. In general, the Spanish factory concentrated on building the more 'basic' cars, as well as commercial vehicles, but it also assembled some of the French Hispanos for wealthier Spanish customers.

The first of the entirely French-conceived types was the H6B of 1919, which had naturally been designed by Birkigt, who now spent most of his time in and around the French capital. The H6B was intended as a fast, luxurious and expensive machine. 80mph was possible, even with the heaviest saloon car coachwork, due to the use of an advanced 6597cc six-cylinder engine, which was effectively half of an intended military V12 aero-engine. The cylinder block was in aluminum, with steel liners, and there was an overhead camshaft with what became famous as typical Hispano-Suiza valve gear. Peak power was 135hp at 2600rpm. This substantial output, and performance, was kept firmly in check by the first successful use of four-wheel brakes with a mechanical servo. This servo was mechanically so elegant, and efficient, that Rolls-Royce soon acquired a licence to use it on their own cars.

Versions of the H6B (or 37.2hp model, as it was sometimes known), performed well

Right and above: 1912 Hispano Suiza.
Below: 1912 Hispano Suiza Alfonso, named after His Most Catholic Majesty Alfonso XIII, the last Bourbon King of Spain. Royal patronage ensured success for the Marque.

in motor sport; the short-wheelbase model was named 'Monza' after a victory at that circuit in 1922. As a result of victories by Dubonnet and Bablot, in the Boillot Cup race at Boulogne, and by Garnier and Boyriven in the Coupe de Boulogne event, a true sportscar version was launched in 1924, known as the Boulogne, or H6C Sport.

Compared with the fist H6B, the Boulogne had an increased cylinder bore (of 110mm in place of 100mm), but the same 140mm stroke, the capacity therefore becoming 7983cc. It had a higher compression ratio and, in the case of highlift camshaft models, more than 200hp was developed.

In 1924 Woolf Barnato (later to become a famous 'Bentley Boy'), drove one for 300 miles at an average speed of 92.2mph, at Brooklands, to gain numerous international endurance records. Another car also won a £5000 wager against Stutz, by averaging 70mph for 24 hours. The top speed of the standard model was about 110mph, and by the standards of the day the handling, braking and road behavior were all superb.

H6 types continued to be built into the 1930s, but were joined, and ultimately replaced, in 1931, by the magnificent French-built V12 models. These were large, fast luxury cars for the very rich, and although a few had open coachwork, most were not sporting cars, despite possessing all the usual virtues of this splendid marque.

Only 15 genuine Boulognes were ever built, nine of them specifically for racing, but many 'normal' H6Bs were fitted with the same type of 8-liter engines, and therefore could achieve nearly everything offered by the Boulognes, though the handling, on the longer wheelbase, was not as good.

As the 1930s progressed, the French factory became increasingly involved in military rearmament, such that production of cars became a sideline. In 1938 it ceased completely, although it continued, in Spain, into the war years.

In total, less than 3000 Hispano-Suizas were built in France, this figure including a few cheaper models made as a result of the takeover of Ballot in 1930. Surprisingly, a few of the more exotic types were made, under licence, by Skoda of Czechoslovakia (in the 'vintage' era), and by an Argentinian firm until 1942.

The Hispano-Suiza is now chiefly remembered as a luxury car for enthusiast owner-drivers, rather than for tycoons leaving the driving chore to their chauffeurs. By these exalted standards, the Alfonso, and Boulogne models were truly exceptional. Like Bugattis, they combined a remarkable blend of engineering excellence and styling harmony.

Marc Birkigt himself finally retired from aero-engine design in 1950, and the French end of his firm then combined with Bugatti in this endeavor.

Right: 1922 Hispano Suiza H6B Labourdette Torpedo.
Below: 1924 Hispano Suiza Tulip Wood H6C.

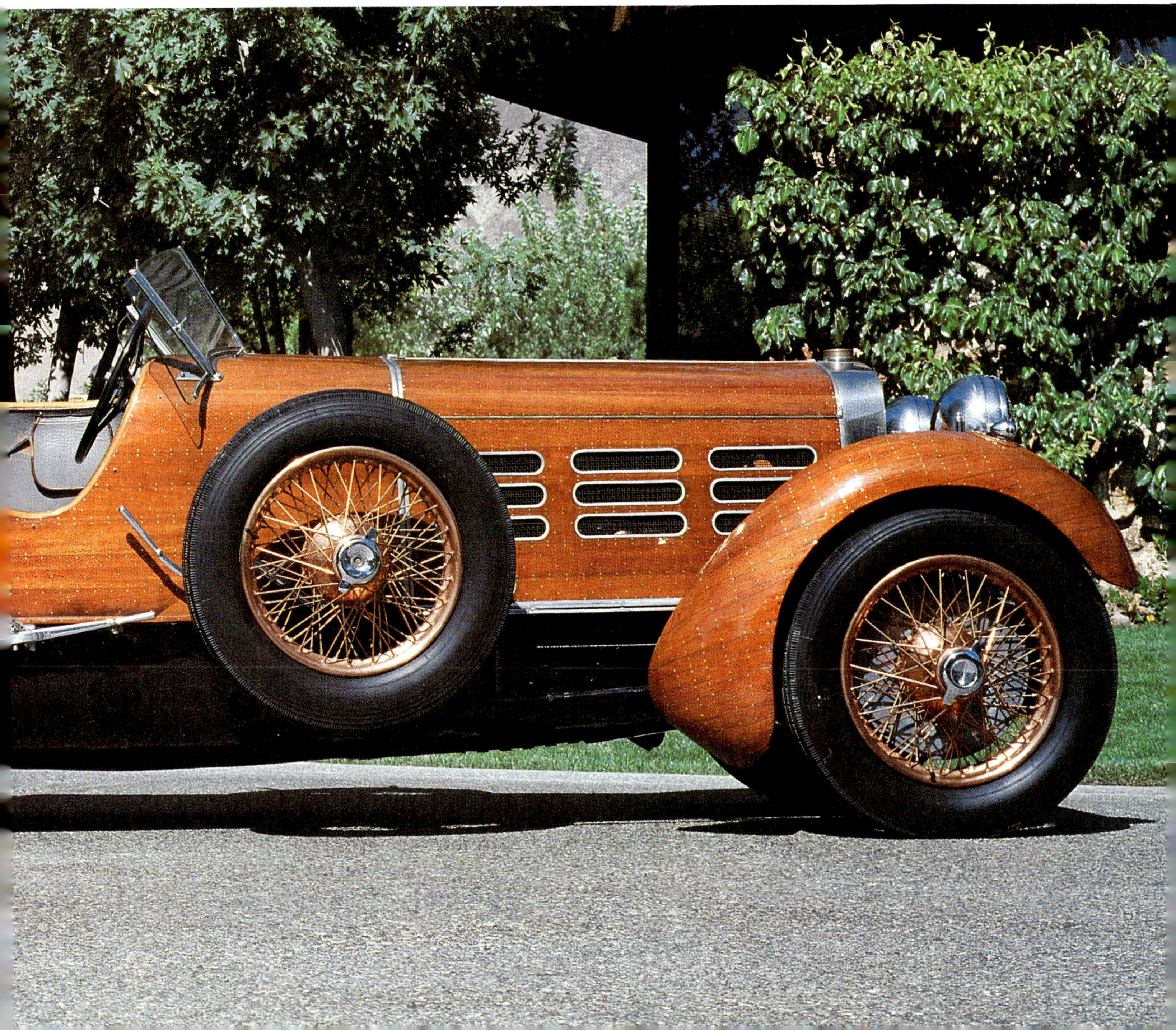

HRG

When Ron Godfrey and Archie Frazer Nash closed down GN production, they went their separate ways. Frazer Nash started building cars carrying his own name, but it was some years before Godfrey went into partnership with Messrs Halford and Robins to produce the first HRG (Halford, Robins, Godfrey) in 1935.

The first cars used the Meadows 1500cc overhead valve engine, which developed 58hp at 4500rpm, and because this was coupled with a compact, 1550lb car, it meant the cars had very sporting performance. Like the Frazer Nash, the HRG had very simple, traditional styling with cycle type wings and virtually no spring movement at all.

The cars had a conventional transmission, using a proprietary Moss gearbox, and the brakes were mostly manufactured by HRG themselves. Production was low, and most cars sold were used in trials, and sporting events. In 1937, Halford himself finished second in the 1500cc class at the Le Mans 24 Hour race, while in 1939 Peter Clark won his class, in the same car.

In 1937 the overhead camshaft 1.5-liter Singer engine was adopted, and at the same time an 1100cc version became available. The 1496cc engined car offered a top speed of about 85mph (from a peak output of 65hp), while the 1100 offered perhaps 75mph.

After World War II, production began again in 1946, with the traditional-style 1100s and 1500s continuing with only detail changes. However, the Aerodynamic model (1500 only) was a real surprise. This used the same chassis as the established cars, which was narrow, but had a full width, all-enveloping, body style, which was not a success. This shell had to be supported from the narrow frame by a series of outriggers, not only at the sides but also at the front and rear. These deteriorated rapidly, and most of the aerodynamics have since been rebodied as 'standard' cars.

The HRG's main postwar competitor was the MG sportscar – the TC at first, the TD from 1950. The MGs were about £200 cheaper, available in quantity, and the TD gave a much softer ride, thanks to the use of coil spring independent front suspension. The 'Hurg,' as it was popularly known, retained a front beam axle on very stiff springs, and while many thought it had an acceptable ride, it appealed to the more masochistic enthusiast.

Competition successes continued to come to the Singer-engined cars. In 1948, for instance, six cars were entered in the French Alpine Rally and, between them, won two team prizes, a Coupe des Alpes, a class win, and three special test awards. In 1949 Peter Clark returned to Le Mans with a

Above: 1947 HRG 1500 cockpit.
Top right: 1947 HRG 1100, pictured at Brunton Hillclimb, 1952.
Bottom right: 1947 HRG 1500 Aerodynamic.
Below: 1937 HRG 1500 photographed at the 1937 Torquay Rally.

LPF 394

131
131
LPH 96

team of three lightened, and special-bodied, cars. Eric Thompson and Jack Fairman won their class, and one of these cars also won its class in the Spa 24 Hour event.

Later developments included the substitution of the Singer SM engine, instead of the obsolete 1500cc engine, and eventually a twin-overhead-camshaft version of this unit was completed. Not only was this engine to have been fitted to an entirely new model (of which three prototypes were completed), but it was to have been offered in a Singer saloon model as well. The prototype HRG had a tubular chassis frame, and disk brakes of HRG's own design. It was to have been sold at a price of £1867, but this was far too high for a 1500cc sportscar of the mid-1950s.

A final HRG prototype appeared in 1965, with a 1.6-liter Vauxhall VX 4/90 engine, but this was purely a one-off, and the company finally closed down in 1966. Total production of the Singer-engined cars was 187, of which 49 had the 1100cc engine, and 138 the 1500 unit.

Right: 1947 HRG 1500 by Real.
Below: 1949 HRG Le Mans Lightweight.

Invicta

Noel Macklin had previously been involved in the Eric Campbell and Silver Hawk motorcar projects before he teamed up with Oliver Lyle in 1925, to make a series of fine sportscars which were to use large and mechanically lazy proprietary engines. Lyle incidentally, was a member of the sugar manufacturing family.

The first Invictas of all used Coventry-Climas six-cylinder engines, but the well-known Meadows six was soon standardized. At first of 2½ liters, but 3 liters from 1925, it quickly earned the marque an excellent reputation. Examples won the RAC's coveted Dewar Trophy on two occasions, and the model became very successful in long-distance sporting reliability trials.

By 1928 these cars had been joined by a 4½-liter engined model, which Invicta hoped would be a competitor to the contemporary Bentley. One subsequent version, the NLC type, was almost as expensive as a Rolls-Royce 20/25hp model, but once the Depression arrived in Europe that put an end to all Invicta's aspirations in the carriage trade.

Instead, a less luxurious 4½-liter 'High Chassis' tourer appeared in time for 1930, and this was soon joined by a new Type S sportscar, usually known as a 'Low Chassis' model. On the 'S' the rear of the chassis frame passed underneath the rear axle, to help give the car a lower center of gravity, and more stable handling. (Even so, *The Autocar*'s Sammy Davis was involved in a well-publicized accident at Brooklands in 1931, when the handling proved to be anything but stable.) Even in basic form, the Meadows engine developed about 120hp, and could propel the car up to nearly 100mph. By 1934, when the last of 77 Type S cars had been built, the power output had risen to 140hp, and a 100mph top speed was guaranteed.

Macklin was keen to prove the 42-liter car's reputation in long-distance touring car events, and the Type S won numerous internal events. Its most famous victory came in 1931 when Donald Healey used one to win the Monte Carlo rally, in spite of having to drive it all the way from the start with a bent chassis, following an accident!

Unfortunately, the Depression, and its financial after-effects, made life very diffi-

Right and below: Invicta 4.5-liter S Type, from the year Donald Healey drove a similar machine to win the Monte Carlo Rally.

cult for Invicta, and production effectively ended in 1935, although there were two subsequent attempts to revive it.

The S's successor has been intended to be a magnificent 4.9-liter twin-overhead-camshaft SS, but only two prototypes of this car were ever built. A similar type of Meadows engine, reduced in size to 3 liters, then turned up after World War II, in the final Invicta, the Black Prince model.

Invicta, in fact, moved out of its Cobham, Surrey, works in 1933, into Chelsea, south-west London, the Cobham plant then being used to make the new Railton car. This was also the inspiration of Macklin, and Reid Railton, and retained the famous Invicta exposed-rivet details on the bonnet panels. It was, however, a much cheaper affair which utilized Terraplane and Hudson (of the USA) chassis and mechanical components, that gave the cars high performance, without any of the true pedigree of an Invicta.

Right: 1931 4.5-liter Invicta.
Below: 1932 4.5-liter Invicta S Type. The low-chassis model went out of production in 1935.

Jaguar

Every Jaguar ever built has had sporting pretensions. Starting from humble beginnings, William Lyons brought style to Swallow sidecars, then to a whole series of special bodies for such mundane chassis as the Austin Seven, the Fiat Tipo 509A, the Standard 9 and the Standard 16. At a time when other companies were struggling even to stay in business, SS, run by Lyons and his partner William Walmsley, flourished, and the next step forward saw Lyons asking Standard not only to supply mechanical components, but to design a new chassis on to which he designed a rakish new body.

The result was the original SS1, and this soon became a range of saloons, coupes and convertibles. The most famous derivative was the two-seater sportscar called the SS90, of 1935, this almost immediately becoming the SS100, first offered with an overhead-valve 2½-liter six, and later a 3½-liter; the first being a converted Standard engine, the 3½-liter being almost entirely an SS design. The SS100 was very fast, for it had a low-slung chassis, a rakish open sports style, with an exposed slab fuel tank, and the roadholding was respectable by the standards of the day.

The name of 'Jaguar' appeared for the first time in the autumn of 1935. The saloons and drop-head coupes, along with the SS100, were all distinguished by unmistakeable styling, which some thought was almost too flamboyant. Yet nobody ever complained about the prices – the 3½-liter S100 sold for £445, could just reach the magic 100mph, and accelerate from rest to 60mph in about 10.5 seconds. Not surprisingly, this performance led several owners to use the cars in competitions. Examples were raced at Brooklands, and hillclimbed at Shelsley Walsh, but the most notable successes were gained in rallies. Tommy Wisdom put up a shattering performance in the Alpine Trial, and Messrs Harrop and Taylor were also outstanding in the RAC Rallies of 1937 and 1938.

In the early postwar years, in the late 1940s, SS100s were also used to good effect, with a young man called Ian Appleyard performing brilliantly in the Alpine and Tulip rallies. The SS100 did not re-enter production after World War II, however, and it was not effectively replaced until 1948.

That year, at the Earls Court Motor Show, Jaguar startled the world by launching the sensational new XK120 Roadster, whose implied top speed of 120mph made it the world's fastest production car of the day. Originally conceived as a limited-production sportscar to prove out the new twin-overhead-camshaft six-cylinder XK engine (which was intended for use in a future range of saloons, to be known as the Mk VII

Below: 1935 3.5-liter SS 100, forerunner of the postwar Jaguar XK Series, and capable of 100mph.

OAU 84

Left: 1951 Jaguar XK 120 Roadster.
Right: Jaguar XK 140 Roadsters.

models), it was such an instant success that more ambitious plans had to be laid, and full body tooling had to be ordered.

The new car set unmatched standards, at the time, for high performance, a supple ride, roadholding, styling, practicality, docility and amazing value for money. The new 3.4-liter engine developed 160hp, and could propel the XK120 up to 60mph in less than 10 seconds. All this was available (after a long wait) for a mere £1298.

The XK120's potential was so startling that many pundits dismissed the claims as exaggerated publicity. To counter this, the company organized a high-speed demonstration run on the Jabbeke highway in Belgium, where all doubts were dispelled, when a standard car achieved an officially timed 126mph (132mph with windscreen removed). To complete the demonstration, the record car was then sedately motored past the assembled journalists, at 10mph in top gear!

There was an ideal opportunity to show off the XK120's all-round prowess during 1949, when the British Racing Driver's Club Silverstone race meeting included an hour-long race for production cars. Jaguar lent three cars to distinguished drivers, and only a tire blow-out to B Bira's car prevented a clean sweep, though Leslie Johnson and Peter Walker finished in first and second places.

Meanwhile, in the rally world Ian Appleyard started to use a white XK120, later to become famous, with the NUB 120 registration number. In the Alpine rallies of 1950, 1951 and 1952, he won Coupes des Alpes for unpenalized runs on each occasion, while he won the Tulip and RAC rallies of 1951. Stirling Moss also enhanced his young reputation considerably when he raced Tommy Wisdom's XK120 in masterly fashion, to win the 1950 Tourist Trophy race in appalling weather conditions. This success was directly responsible for giving Moss a place in the Jaguar 'works' team.

Above: Jaguar XK 140 grille.
Left: 1952 Jaguar XK 120 Roadster. It was only because Sir William Lyons thought there should be a sportscar in the series that they were built.

The XK120 Fixed Head Coupe was launched at the 1951 Geneva Motor Show, having a shape rather reminiscent of certain late-1930s Bugattis, and of the one-off SS100 Coupe seen in 1938. Because it was rather more civilized than the Roadster, it made an excellent long-distance touring car. The interior was altogether better trimmed, and there was a veneer-trimmed dashboard, and wind-up door windows. A Special Equipment model was offered, with a 180hp engine, these models being known, in the USA, as XK120M.

In 1950 the factory entered a team of three XK120s in the Le Mans 24 Hour race, not seeking an outright win, but seeking to gain experience, and a respectable finish.

Above: Jaguar XK 140 fixed Head Coupe with now-traditional sportscar fastback lines.

Two cars finished, 12th and 15th, and from this it was felt that a specially-designed car, using many components from the production model, might stand a good chance of success in 1951.

The result was the multi-tube frame XK120C, more usually known as the C-Type, which achieved exactly what was intended, for the Walker/Whitehead car won the 1951 race outright. In 1952 all the team cars retired, following ill-judged tinkering with cooling systems and the body shape, but in 1953 they returned triumphantly. With more powerful (220hp) engines than ever before, and the new-fangled Dunlop disk brakes, they were dominant – taking first, second and fourth places.

In the same year the third variation on the XK120 theme was introduced, this being the Drophead Coupe. it had all the refinements of the Fixed Head model, but was fitted with a fully-trimmed, convertible hood. In production terms this was the rarest of the 120s – 1765 beng produced, compared with 7612 Roadsters and 2678 Fixed Heads.

In October 1954, the XK140 took over directly from the 120, though remaining closely based on that design. Changes included the moving forward of the engine in the chassis, to allow the cockpit to be enlarged, the increasing of power to 190hp, and the fitment of rack-and-pinion steering. The price of the Roadster had risen to £1598. External style changes included the fitment of larger bumpers, a cast grille, and extra chrome decoration on bonnet and bootlid. The Drophead and Fixed Head models now had space for two tiny occasional seats, suitable only for carrying children.

On the competiton side, 1954 had also seen the arrival of the now-legendary D-Type, perhaps the most famous of all '50s racing sportscars. The 'D,' like the 'C' before it, had been designed with only one aim in mind – that of winning the Le Mans race, and reaping all the publicity and prestige which would follow. The car had a central monocoque, using aircraft principles, and the shape had been refined by wind-tunnel testing. Efficient aerodynamics were of great importance because the Le Mans circuit had a four-mile long flat out straight. Even at this time the lap record was higher than 110mph.

By this time the competition XK engines had modified heads, three Weber carburetors, and produced 250hp at 6000rpm. Dry sump lubrication systems had been installed, and this allowed the engines to be lowered, with benefits to the bonnet profile, and to the car's center of gravity. Independent front suspension was by longitudinal torsion bars and wishbones, the rigid rear axle was located by trailing arms and tranverse torsion bars, and there were Dunlop disk brakes all round.

The D-Type's first Le Mans race was frustrating for Jaguar. Although the cars performed well, they were hampered by fuel starvation, caused by the presence of sand in the fuel. After this was diagnosed and rectified, the Rolt/Hamilton car made spirited efforts to retake the lead, and in spite of being forced off the track at one point by a slower car, they finished second, just 105 seconds behind the winning 4.9-liter Ferrari. Revenge followed within weeks, when the D-Types finished first and second in the Rheims 12-Hour race.

For 1955 the 'works' D-Types had even better aerodynamics, with a longer nose, smoother tail fin behind the driver's seat, and more power. A new 'wide-angle' head with larger valves was used, and power was up to 275hp. Under the skin, there were constructional changes, the front sub-frame now being bolted, rather than welded, to the central monocoque.

The D-Type, driven by Mike Hawthorn and Ivor Bueb, won the Le Mans race of that year, though it was a sad occasion following Levegh's disastrous crash in a Mercedes-Benz 300SLR, and the subsequent with-

drawal of the German team. During a fantastic race for the lead in the early hours, Hawthorn had set a new lap record at 122.39mph.

The following year's event saw two of the fuel-injected team cars eliminated by an accident on the second lap, and the third delayed by mechanical bothers, but a private team, Ecurie Ecosse, rescued the day by winning in one of their own D-Types. After 1956, Jaguar withdrew the 'works' team, and the 1957 Le Mans race was won, once again, by an Ecurie Ecosse car.

In the meantime, Jaguar had put the short-nose 250hp D-Type into limited production,

Right: D Type Jaguar race car.
Below: C Type Jaguar. Flowing lines of the E-Type road cars are already evident.

but this was all over by 1956, and a few thinly-disguised cars, with sketchy road equipment became XK SSs in 1957.

In 1957, too, the XK theme was updated further, by the introduction of the XK150. This used essentially the same chassis as the XK140, though Dunlop disk brakes had been adopted, all round, but the body style had been much-changed, with a wider bonnet, and more cockpit space. There was improved engine bay access, and the wing line profile, view from the side, was now almost straight, instead of being rather swooping. Instead of a vee-profile, there was now a one-piece semi-wrap-round windscreen. Almost every XK150 was sold with the 210hp 'special equipment' engine, through the less powerful version with a 190hp tune was still available.

The car was beginning to look, and feel, its age, but still remained tremendous value for money, compared with its rivals. The XK150 DHC, for instance, cost £1783, com-

Top left: 1954 D Type Jaguar, the car which put the Jaguar name into racing folklore with three wins at Le Mans.
Bottom left: 1957 Jaguar XK SS.
Left: Lister-Jaguar still in competition use.
Below: 1959 Jaguar XK 150S fixed Head Coupe.

14
Pace Petroleum

750 DUK

Above left: 1961 Jaguar E-Type racer.
Below left: Jaguar E-Type Convertible was the definitive British sportscar for over a decade.
Above: The Jaguar Lightweight E-Type which Phil Hill raced in the 1963 24 Heures du Mans.

pared with £2889 for the Aston Martin DB2/4, £4201 for the equivalent BMW, and £4651 for the Mercedes-Benz 300SL. At first there were only drop-head and fixed-head types, but one year later the two-seater Roadster (with wind-up windows in the doors) was also announced, along with a 250hp triple-SU carburetor 'S' engine tune option. In 1959, not only was a 220hp 3781cc engine made available for all derivatives, but also a 265hp 3.8-liter S tune as an option.

Just as in 1948, so in 1961 did the evolution of the sportscar take a great leap forward, when Jaguar unveiled its most famous sportscar, the E-Type. Based, in many ways, on the construction of the famed D-Type, the E-Type was a great leap forward, and caught the imagination of press and enthusiasts alike. Known as the XKE in the United States, the car was an instant success, and earned millions of dollars in Jaguar's most important export market.

The E-Type's construction was based around a pressed-steel monocoque, with a bolt-on front tubular sub-frame carrying the engine and front suspension. Under the rear of the monocoque there was a 'cage' type of sub-frame supporting the final drive, and coil spring independent rear suspension, a new departure for Jaguar. This had already been tried, briefly, on the E2A prototype which raced at Le Mans in 1960, and helped provide a supple ride to this, and also to future Jaguar saloons.

The engine used was a 3.8 'S' version of the XK unit, the imroved aerodynamics and lighter weight both helping the new car to a top speed of 150mph, and 0-60mph acceleration in 6.9 seconds. Two body styles were offered, an open two-seater, and a fastback fixed-head coupe, with opening rear window. These cars sold in the UK for £1830 and £1954.

Like the XK120 before it, the E-Type won its first race, this time driven by Graham Hill. However, it did not have a very successful international racing career, because racing sportscars were becoming more and more specialized all the time, and the E-Type could not match the latest Ferraris.

To attempt to rectify this, Jaguar built a dozen 'Lightweight' E-Types, some with an aluminum monocoque and aluminum cylinder block for the engine. This unit, in fuel injection form, produced more than 320hp – twice that of the original XK120 unit of 1948. In spite of being driven by great racing 'names' like Graham Hill, Roy Salvadori and Jackie Stewart, these cars were only partially successful, over a short period.

In 1964 the E-Type was given the latest 4235cc version of the XK engine, in place of the 3.8-liter unit, and though the peak power output remained the same there was significantly improved torque. At the same time a new all-synchromesh gearbox replaced the old-style box, which had always been noted for 'slow,' rather weak, synchromesh. Production figures for the 3.8 had been 7827 Roadsters, and 7669 Fixed Head Coupes, but the 4.2-liter was a little more popular, selling 9548 and 7770 respectively.

In 1966 the E-Type range expanded, to provide a nine-inch longer wheelbase fixed-head coupe, with more vertical windscreen, and 2+2 seating, for the 'sporting family man.' Two years later all models underwent modifications, to keep them up with the latest USA safety regulations, this including the deleting of front headlamp covers, the fitment of larger tail lamps, and the raising of bumpers. These, and other details, made the E-Type into a Series 2 model.

By the end of the 1960s, however, the E-Type needed another boost, and this came with a really dramatic technical leap forward, when the fabulous single-cam-per-bank V12 of 5343cc was launched. Originally designed as a four-cam racing engine, and then intended for the big XJ saloons, it arrived first in the sportscars with a 304hp power output and massive torque. The E-Type's performance had been slipping back in recent years, due to aerodynamics changes, and emissions legislation, but at a stroke its status as a Supercar was restored.

The range of bodies was optimized for this, the Series III car, for the longer-wheelbase underpan was used, and there were two-seater Roadster, and 2+2 Fixed Head Coupe versions of this. The last of all were built in the winter of 1974/75, there being 7990 Roadsters and 7297 Fixed-heads, respectively.

In the autumn of 1975, therefore, the V12 E-Type was replaced, though not directly, by the XJ-S Coupe. This was an entirely different type of sporting car from the E-Type, having a pure monocoque construction, based on the underpan of the 12-cylinder XJ12 saloon, and being more of a high-speed 'executive express' rather than a sportscar. Nevertheless, it was technically superb, and before long every XJ-S was being built with automatic transmission. Front and rear suspension were independent, by coil springs and wishbones at the front, by a modified type of E-Type/XJ12 system at the rear, and there was 2+2 seating. The concept was significantly improved in mid-1981, when the more efficient and more economical 'HE' engine was introduced, the car then becoming known as XJ-S HE.

From 1982 the XJ-S began to make its name in production touring car racing, when Tom Walkinshaw prepared two cars for the European Touring Car Championship. Initial success included victory in the Tourist Trophy, after which Jaguar gave official support for 1983 and 1984. The result was that Jaguar scored five outright victories (to six from BMW) in 1983, then with a full three-car team went on to win the Manufacturer's and the Driver's (Tom Walkinshaw himself) series in 1984.

The latest variation on the XJ-S theme came in the autumn of 1983, when Jaguar not only launched a cabriolet version of the XJ-S, which featured a fixed roll-over bar above the seats, but they also put a completely new 3.6-liter twin-cam six-cylinder engine of 225hp, linked to a five-speed Getrag gearbox, on sale. The engine is clearly intended for other Jaguar models in the future.

Right: Interior, 1961 Jaguar E-Type S2 4.2-liter.
Below: 1961 Jaguar E-Type 3.8-liter Roadster.
Top right: 1985 Jaguar XJ–S C 3.6-liter Cabriolet with the new six-cylinder engine.

B103 YMK

Lagonda

Although the name Lagonda has its roots in the USA, it has always been that of a British automobile company. 'Lagonda' was in Ohio, where the founder of the company, Wilbur Gunn, first lived. At one time he was an opera singer, but he came to England in 1898 to further his career. Shortly, he became involved in steam-powered boats, and the transition to road vehicles followed rapidly. The very first Lagondas were three-wheelers, but the marque soon developed a high-quality small car, which performed extremely well in reliability trials up to the 1914 period.

During the late 1920s, however, Lagonda reversed all their previous policies, and began to concentrate their efforts on the production of high-quality sportscars. The first of these was the Speed Model, also known as the 2-liter, and was announced in 1927. This car had a four-cylinder 1954cc engine with twin overhead camshaft valve gear, and was available with or without a supercharger. Like most such 'vintage' sportscars, it had a rather flexible separate chassis frame, beam axles front and rear, with half-elliptic leaf springs.

Below: 1928 2-liter Lagonda High Chassis Speed Model.

The standard car had a good performance for the period – it could do 70mph in third gear, and 80mph flat out, while the supercharged version was good for 80mph in third and more than 90mph in top. The chassis sold for £530, the tourer for £675, and the Weymann-bodied saloon for £750, which were high prices for the 1927 period.

In 1928, the 3-liter model was announced. This had a six-cylinder engine of 2931cc (which was enlarged to 3181cc for 1933), and used a lengthened version of the 2-liter's chassis frame and suspensions. By 1931, too, this had also evolved into the Selector Special, so called because of its Maybach pre-selector transmission. This unique design endowed the car with two different sets of four ratios. Since it was also possible to use no fewer than four reverse ratios, it was advertised as the '12 speed Lagonda,' which rather confused the public. The Weymann bodied car sold for £1065, and the open version for £975.

The 4½-liter M45 model was announced in 1933, and was to become the best known and most successful of the first generation of sporting Lagondas. This model was fitted with a six-cylinder overhead valve engine from Meadows (the same basic unit as that used by Invicta, except that the Lagonda unit produced more power). A more highly-developed model named the M45R was produced, which developed 130hp at 3800rpm. The saloon version had a top speed of 90mph, and sold for £950, very reasonable compared with the £1380 asked for a 3½-liter Bentley.

In 1934 and 1935 the 4½-liter was successful in motor racing, with 'works' cars being prepared by Messrs Fox and Nicholl. A tourer came fourth in the 1934 Tourist Trophy, while Hindmarsh and Fontes won the 1935 Le Mans race at 77mph. This was to be the only victory for a British car, at Le Mans, between the Bentley era which ended in 1930, and the first Jaguar victory of 1951.

Commercial disaster followed, however, for the Receiver had been called in just before the Le Mans race, and no publicity benefit could be taken from this great win. Eventually the business was bought by Alan Good, who paid just £67,000, plus an extra £4000 for stock. Production was started up again, but only of the 4½-liter models, and the LG45 of September was an improved version of the original. Good had taken on W O Bentley (now clear of his commitments to Rolls-Royce), and his first important contribution was to influence the latest engine, complete with a cross-flow cylinder head.

At the same time, body engineer Frank Feeley was given his head, to design an

sion, by wishbones and longitudinal torsion bars, for the very first time on any Lagonda, while the rigid rear axle rode on half-elliptic leaf springs.

Most V12 chassis were fitted with big and graceful saloon car bodies, but there were a few sportscars, and the factory also built competition models for the 1939 Le Mans race. They were splendid, and expensive, cars, priced at £1285 for a tourer, and £1585 for the coupe. It took time to get the V12 model into series production, but just when reliability was becoming assured, World War II broke out, and the car was killed off. It was never to be revived.

The postwar Lagonda model, developed before the end of hostilities, was prematurely announced in 1945, and was at one time to have been called a Lagonda-Bentley. Rolls-Royce, however, insisting that they owned the rights to the 'Bentley' marque, reacted violently, and so it reverted to its basic 'Lagonda' title. Unfortunately, the capital needed to put the new model into production could not be found, so it had to wait until after the company had been purchased by David Brown in 1947, and began to merge the business with Aston Martin, which he also bought in the same year.

The new car had a cruciform-style chassis frame, all-independent suspension, and the engine was a 2580cc straight six-cylinder unit, with twin overhead camshaft valve gear. The Lagonda saloons and coupes produced up to 1953 were Grand Touring, rather than sports models, but the

Above: 1930 Lagonda 3-liter Tourer.
Right: 1931 Lagonda 2-liter Speed model. WO Bentley later joined the company and designed a 100mph V12 engine to replace the Meadows power units.

eye-catching sportscar body for a new derivative, the LG45 Rapide, the result being reminiscent of the Mercedes 540K Models, complete with huge outside exhaust pipes, and a lengthy bonnet. W O Bentley, it is said, did not care for this, for he was too involved in a new engine project to object.

The new engine, as inspired by Bentley, was shown in prototype form in 1936, but was not ready for sale until 1938. The intention had been to achieve the same standards as he had already done with the 8-liter Bentley of 1930 – namely to 'propel a large closed bodywork in complete silence at over 100mph.' It was no longer necessary to use such a large engine, and as Bentley had already seen the designs of the new Rolls-Royce Phantom III V12 unit before he had left that concern, this may have influenced him to choose a V12 layout for the new Lagonda unit.

With the help of engineers poached from Rolls-Royce, 'WO' produced a 4480cc V12, with a single overhead camshaft valve gear to each bank, the result being a power output of 157hp at 5000rpm. A competition version developed 207hp at 5500rpm. The chassis featured independent front suspen-

APH 13

CPE189

engine soon found a home in the Aston Martin DB2, DB2/4 and DB Mk III models which sold so well up to 1959.

A completely new racing sports Lagonda was announced by the Aston-Martin-Lagonda company in 1954, specifically aimed at the Le Mans sportscar race. In all the best Lagonda traditions, this had a 4½-liter V12 engine, but this was fearsomely complicated, with its four overhead camshafts, its 24 spark plugs, and its three four-choke downdraught carburetors. The body style was similar to that of the six-cylinder engined Aston Martin DB3S models, but the cars were much heavier, at 2350lb, and needed all of their 300hp to make them competitive. All in all, four cars were built by the factory, but they were retired after achieving no successes.

Since then, the Lagonda name has taken a back seat, in sporting terms, to the Aston Martin marque, though it now graces the most striking car to be built at Newport Pagnell. After Sir David Brown sold out in 1972, the company changed hands again in 1975, and the new owners took a bold decision, to promote a large four-door four-seater saloon, with sharp-edged styling and advanced electronics, to compete in the rarified Rolls-Royce price bracket. This car, first delivered from 1978, and current in 1985, used an Aston Martin four-cam V8 engine of 5340cc, and was a Grand Tourer in the true sense of that phrase. Its 1985 price was £65,999, compared with the £58,037 asked for the Rolls-Royce Silver Spirit.

Above left: 1933 Lagonda 3-liter Tourer.
Above right: 1934 Lagonda 16/80 2-liter.
Below and *below left:* 1935 Lagonda M45 Rapide 4.5-liter Tourer; a Meadows-engined Lagonda won the Le Mans 24 hours in 1935.

Lago Talbot

The Lago Talbot marque was created in 1935, when Major Tony Lago purchased the French branch of the bankrupted Sunbeam-Talbot-Darracq combine. However, the story really begins much earlier, at the end of the 19th century, when Adolphe Clément, a French industrialist, began to produce cars. He was interested in exporting his cars to the UK, and soon became associated with the Earl of Shrewsbury, and Talbot, who called the imported cars Clement-Talbots. This was such a success that local assembly, in London, began in 1904. Soon a sporting reputation was gained, and in 1912 a special version of the 4.5-liter model, driven by Percy Lambert, became the first ever machine to achieve 100 miles in one hour, on the Brooklands track.

In 1919, however, the Earl of Shrewsbury sold his business to Darracq, who then went on to form the first major international alliance, to join up with Sunbeam, and set up the S-T-D combine in 1920. Things became increasingly confusing in 1922, when the French concern started to sell its own cars as Talbot-Darracqs, and then as Talbots. The British arm continued to produce Talbot and Sunbeam cars, in separate factories, until they were both taken over, after the liquidation, by Rootes in 1935.

Shortly after Tony Lago bought the French concern, he introduced a new range of six-cylinder cars, with overhead valves and 3996cc, which he called the Baby 4-liter. A sports racing version, called the Lago Special, was also produced, this having a 165hp engine, and a top speed in excess of 110mph. These cars were very successful in competitions, for in 1937 they took the first three places in the Montlhéry sportscar Grand Prix, and in the same year Comotti won the Tourist Trophy race at Donington Park.

The touring cars were well matched to the specialist body styles of the period, but somehow the cars were not considered as chic as the Delahayes. In 1938, however, the engine size was increased to 4.5 liters, and racing successes continued, with a win in the Paris 12 hour race.

Great things were promised for 1939, when a 3-liter V16 engine, no less, was announced, but because of the imminent outbreak of war, nothing more was heard of this, and it was not seen again.

After the war, Lago Talbot returned to private car production in 1947, with improved versions of the 4.5-liter car. A great aid to the marque's image was given by the production of a single-seater Grand Prix car (4.5 liters, unsupercharged, fought in the same formula with the 1½-liter supercharged models), and these had success by being much more fuel-efficient, and not needing to make mid-race refuelling stops.

In 1947, Rosier won the Albi race, while Chiron won the French GP, and at Comminges the big cars were first, second and third. Although they did not often beat the supercharged Alfa Romeos to win, they were usually reliable, and this pointed the way forward, for Ferrari to start beating Alfa Romeo, in similar fashion, from 1950.

In 1948 a new engine with twin high (but not overhead) camshafts was introduced, and power rose from 180hp to 280hp. A two-seater sports version of the racing car won the 1950 Le Mans race, driven by Rosier, and Pierre Levegh so nearly repeated the trick in the 1952 event, driving single-handed for more than 22 hours, before the engine blew up. There was no success in 1953, in spite of the arrival of a new and lightweight car, and by this time the engine could be developed no further. A new racing sportscar, fitted with a Maserati engine, was also tried, but again without success.

The Lago Talbot road car of 1947 used a 170hp engine, bodies normally being supplied by specialist coachbuilders in a variety of styles, mostly looking very traditional. From 1949, too, there was the Talbot 'Baby,' which had a 118hp 2.7-liter four-cylinder engine, having cross-pushrod valve gear, this being available as a drop head coupe or a saloon. Peak production was 433 cars, in total, in 1950, but the following year it fell to a mere 80 units.

In 1954, production was concentrated on

Below: 1938 Talbot Lago 4.5-liter Cabriolet. *Right:* 1949 Talbot Lago Grand Sport Le Mans.

Above and right: 1959 Talbot Lago America 2.5-liter. In this year Talbot-Darracq, all that was left of S-T-D, was absorbed by Simca. Tony Lago died the following year.

a new car, which had a four-cylinder, 2476cc engine, in a smooth sports coupe body. It was a modern-looking car, with 115hp, and a maximum speed of 120mph, but it was also the very last to be produced with an engine of Talbot design; only about 70 of this model were ever produced.

By now the company was in serious trouble, but another new model was produced in an attempt to save the concern. This was a small, good-looking, GT coupe, powered by the well-known BMW light-alloy V8 engine of 2580cc. The model itself was dubbed the 'America,' for it was in that market that Lago thought salvation would come.

Compared with the last Talbot-engined car, the America had 10 percent more power, it had a top speed of 124mph, and could reach 60mph from rest in 10.6 seconds. The design came from Carlo Delaisse, who also produced an attractive body style for the car, rather similar to that of the Talbot-powered coupe. This was actually produced by Letourneur et Marchand, and was notable for having a fiberglass roof, and plastic rear window. There was a modernized type of Talbot grille, to give at least some historic throwback. The chassis design was unadventurous, for there was a tubular ladder-type frame, independent front suspension featuring a transverse leaf spring, and a rigid rear axle.

The America, which was the only Talbot ever to have left hand drive, was competently designed, but far too late, and only 12 cars were built before the company was forced into liquidation.

In 1958 Tony Lago reluctantly let his company be absorbed by Simca, and the last cars looking like Talbots actually used the venerable side-valve V8 engine, of Ford heritage, from the Simca Vedette. Thus, through the purchase of Simca by Chrysler, and of the Rootes Group at a later date, Chrysler quite unwittingly brought two of the S-T-D marques back under common ownership. The name of Talbot, of course, is now applied to family cars produced by the ex-Chrysler, ex-Rootes company (now Peugeot-owned!) in the UK.

Lamborghini

When Ferruccio Lamborghini owned a Ferrari of his own, he was once reputedly snubbed by Enzo Ferrari himself, and thenceforth decided to start up a car manufacturing company in competition to him. He had already made his fortune by manufacturing tractors, and industrial heating equipment, so could well afford to see his automotive fantasies translated into expensive reality. He had always owned fast cars, and had competed in the 1948 Mille Miglia, so his aim was not just to build exotic cars, but to create a new breed of Supercar.

To start with, he commissioned a brand-new factory at Sant'Agata Bolognese, just 20 miles from the Ferrari plant at Maranello, while at the same time he hired Giotto Bizzarrini to design an engine, and Gianpaolo Dallara to look after chassis design. Both men had previous Ferrari design experience.

The first Lamborghini road cars – the 350GT and the 400GT – were relatively conventional front-engined cars, distinguished by the magnificent four-cam V12 engine (of 3½ liters and 4 liters respectively). The marque's reputation, however, was suddenly transformed with the unexpected launch of the amazing mid-engined Miura.

This had originally been conceived, by the designers, as a racing sportscar, but Lamborghini himself would have none of it, and demanded that the car should be tamed for road use. Not surprisingly, it had a mid-engined layout, but the real advance was that the long V12 engine had been mounted transversely across the frame, the engine block being combined with the five-speed all-synchromesh transmission.

With an outstandingly beautiful body style by Bertone, and a name taken up from that of a particularly fearsome breed of Spanish fighting bull, the first complete Miura was at the Geneva Motor Show of 1966. Although it was originally only intended for a very short production run, it received a flood of orders, and was destined to become *the* Supercar of the 1960s.

Apart from its general body lines, the Miura was also distinctive because of the way its headlamps lay flush with the bonnet panel, facing the sky when not in use, for the 'eyelashes' below them, and for the slatted rear window. The engine of the original model produced 350hp at 7000rpm, but the more powerful Miura 400S produced 370hp at 7000rpm, and this edged up even more for the final SV version.

The original P400 was somewhat underdeveloped, and the P400S of 1968 was a much more refined car, with a top speed of nearly 170mph, and heavy fuel consumption. All in all, 763 Miuras were built in eight years, to 1973.

If the smaller, V8 engined Urraco is discounted as a sportscar, because of its 2+2 seating and – by Lamborghini standards – less than Supercar performance, then the Countach is the next Lamborghini truly to qualify. The name comes from a Piedmontese exclamation of amazement

All pictures: Feruccio Lamborghini's Miura P400 coupe (this one from 1966) established him as a major sportscar manufacturer.

6th May 1935

which, roughly translated, could be 'Cor' or 'Crikey,' and was an ideal name for the Miura's replacement, first seen in prototype form at the Geneva Motor Show of 1971. Quite simply, the designer's brief was to produce the ultimate no-compromise high-performance road car. Cost was barely a consideration, and the result was dramatic, to say the least, for Bertone's style included doors which did not swing outward, but upward and forward from low-mounted front hinges.

The chassis was a very complicated multi-tubular structure, in which the famous V12 unit was mounted lengthwise behind the cockpit. However, the five-speed transmission was ahead of the engine, protruding into the cockpit, and drive to the rear wheels was by a propeller shaft, through a tube in the engine sump, to the final drive.

As a consequence, the driving position was a long way forward, but the car's center of gravity was low, and there was a 43/57 percent front/rear weight distribution. There was even space for some luggage behind the engine itself. The power unit, originally of 3929cc/375hp, was eventually enlarged to 4754cc with no more power, but better torque, in 1982. A rear spoiler wing was optional, and top speed was at least 180mph.

Above and below: The original Lamborghini Countach. *Left:* The 1985 Lamborghini Countach QV (Quattrovalvole) represented a considerable performance improvement and is the fastest road car to date.

Lancia

With only one exception, pre-1939 Lancias do not qualify for the title of 'sportscar,' but that exception was quite a remarkable car. This was the Lambda model, which was generally agreed to be the first design to feature a simple form of unitary body/chassis construction. In this case it is said that Lancia's inspiration came from studying the construction of boats, while he was taking a cruise. The Lambda was also the very first large car to have independent front suspension, worked by a system of coil springs on sliding pillars, all in enclosed cylinders, and a type of hydraulic shock absorber.

The Lambda was launched at the Paris and London shows of autumn 1922, then produced in nine series until 1931. The engine was a narrow angle (only 13 degrees) V4, with staggered bores, and was so compact that there was also space for the transmission to be mounted under the bonnet. Initially the engine capacity was 2170cc, and produced 49hp, but from the third series it was enlarged to 2370cc/59hp, which increased top speeds from 70mph to 77mph. The fifth series had an even larger engine, of 2570cc and 69hp, which pushed up the top speed to nearly 80mph.

The floor of the passenger compartment was so low that a propeller shaft tunnel had to be incorporated. Even though the Lambda was not at first intended to be a sportscar, it had sporting characteristics. Some 13,000 examples were built, and the car has justly been labelled as a landmark in automotive design.

The firm's founder, Vincenzo Lancia, died in 1937, at the age of 56, but left behind, as his legacy, a very advanced small saloon model, known as the Aprilia, while the company remained in the control of his son, Gianni. It was not until after World War II, however, that Vittorio Jano was brought in (from Alfa Romeo), to become Lancia's design and research chief, and under him a new V6 engine was developed.

The original unit was a 1754cc 60-degree unit, producing 56hp at 4000rpm, and was installed in a new saloon car called the Aurelia B10. The first examples had pleasant, but undistinguished saloon styled on a unit-construction basis, but were mechanically quite special because they had a rear-mounted transmission, in unit with the final drive, inboard rear drum brakes, and independent rear suspension. It was during the Aurelia period that the company's policy toward motor sport changed, for Gianni was adamant that his company could benefit from success; accordingly, much effort went into increasing engine power, and vehicle performance.

The Aurelia saloon had been launched in 1950, but a year later the Pininfarina-styled fastback GT coupe, the B20, was also put on sale. This was a very attractive car, setting many standards for the future, and originally had a 1991cc/80hp V6 engine. GTs enjoyed wide competition success, with a second in the 1951 Mille Miglia, 12th place at Le Mans in the same year, and third in the Mille Miglia of 1952.

Later series of B20s were fitted with De Dion rear suspension, and became more powerful, with 2451cc and 118hp at 5000rpm from the third series. It was cars of this type which won the Liège-Rome-Liège rally in 1953, the Monte Carlo rally of 1954, and the Targa Florio in the same year.

Lancia also developed an open version of this design, also styled by Pininfarina, called the Aurelia Spyder, the first variety having rather ungainly wrap-around windscreens, but the fifth and sixth series cars having a more conventional screen and wind-up windows.

In the early and mid-1950s, Lancia became more and more involved in motor sport, first with racing sportscars, and finally with a Grand Prix F1 design, all to the design of Jano. The first sports racer was the D20, which had a 60-degree four-cam V6 engine of 2962cc, which produced 217hp. This car had a tubular chassis frame, and independent suspension all round, though later varieties had de Dion rear ends. Later the D24 was produced, with a larger, 3300cc engine, and it distinguished itself by taking the first three places in the Mexican road race, the Carrera Panamericana, piloted by such illustrious 'names' as Fangio, Taruffi and Castellotti. The same car also tasted victory in the Tour of Sicily, the Targa Florio, and – driven by Alberto Ascari – the 1954 Mille Miglia.

Below left: 1928 Lancia Lambda 2.4-liter, this from the 3rd series to bear the name.
Below: 1953 Lancia D.24 Carrera.

Development of the D50 Grand Prix single-seater of 1954/55 was the last financial straw, problems ensued, and Gianni Lancia was obliged to hand over the cars to Ferrari, and to sell his company to the Pesenti business interests.

In the late 1950s and the 1960s, Lancia relied for their survival on a series of technically advanced, but mundane looking, touring cars – notably the Appia, Flavia and Fulvia, and although sporting derivatives of each were produced, none qualified as an out-and-out sportscar. Nevertheless, the lightweight Fulvia Coupes were outstanding rally cars, and the Zagato-bodied equivalents were unmistakeable, with excellent performance and roadholding.

In 1969, Lancia, which had been losing money fast, was taken over by Fiat, who also controlled Ferrari by this time, and it was this association which led to the Ferrari Dino 246GT's engine and transmission being used to power the futuristic Stratos. This was designed purely with rallying in mind, with much work credited to Gianpaolo Dallara (ex-Lamborghini, ex de Tomaso), and the styling and structural manufacture by Bertone. As used in the Stratos, the engine was transversely mounted, behind the seats, and produced 190hp at 7000rpm, from 2418cc, driving the rear wheels through a Ferrari Dino five-speed gearbox. Front and rear body sections could be swung completely up, and out of the way, to give access to the all-independent suspension, and four-wheel disk brakes.

During 1974 and 1975, approximately 500 Stratos were built, so that the car could be homologated as a Group 4 competition machine, and once approved it dominated rallying more than any other car had ever done. The 'works' team won three consecutive World rally championships – 1974, 1975 and 1976 – and in 1978, Stratos cars won 13 national and international events, all round the world.

In the meantime, Fiat had produced a mid-engined sports coupe which they then donated to Lancia for production; this was the unit-construction Monte Carlo, with coachwork by Pininfarina, which used a

Above: The 1952 Lancia Aurelia B.20 2.5-liter coupe.

transverse mid-mounted twin-cam Fiat four-cylinder engine of 1.6, 1.8, or 2.0 liters, with integral five-speed transmission. In the USA this car was known as the Scorpion, and in all markets it was sold with a fixed roof, or with a roll-back 'convertible' top. The last of these cars was built in 1983/84, after a two year lay-off in the late 1970s to sort out handling and quality problems.

For rallying purposes, however, the Stratos was succeeded by the Lancia Rally coupe, which looked rather like the Monte Carlo, but was almost entirely special except for using that car's center section. Only 200 were produced, with tubular front and rear sub-frames, a longitudinal instead of transverse engine mounting, and a supercharged (not turbocharged) engine.

The first car was shown at the end of 1981, the 200-off run was completed by mid-1982, and the car started winning later in the year. In 1983 it distinguished itself by winning the Monte Carlo rally, and going on to take the World Rally Championship for Makes. In road-car tune, the 1995cc 16-valve engine developed 205hp, with up to 325hp at 8000rpm available in fully-tuned form.

Above right and right: The Lancia Stratos (this is 1976 model) represented the company's major venture into the international World Rally Championship.
Below: 1955 Lancia Aurelia B.20 coupe.

SGS 961W

Lotus

The late Colin Chapman was one of the greatest innovators of motorcar design. He began by designing and building specials based on the ubiquitous Austin Seven, and eventually rose to controlling a world-famous Grand Prix racing team. Every road car produced by Lotus has received technical acclaim.

The very first 'Lotus' car was built in 1947/48, while Chapman was still studying for his engineering degree at London University, and this was followed by other trials specials, and club racing sportscars built to comply with the regulations of the 750 Motor Club. These were so successful that Chapman was approached by other enthusiasts, who wanted replicas, so soon Chapman found himself in the car-manufacturing business, with the aid of £25 loaned to him by his girl friend Hazel, whom he later married.

The first 'production' Lotus was the Mark 6 model, the first of many such cars to have a multi-tube space frame chassis enclosing the engine and transmission, with soft independent front suspension, and very light weight. Then, and later, his philosophy was that no item should be in any way superfluous, or over-strong, for this simply added unnecessary weight to the machine.

This basic frame was clothed by stressed aluminum body panels, and the assembly weighed just 90lb. At first all Lotuses were 'kit' cars, for the customer to complete in his own garage. The Mk 6 had a Ford back axle, cable-operated Girling brakes also as used by Ford, and – at first – a modified 1.5-liter Ford Consul engine.

When fitted with race-tuned MG TD/TF engines, the Mk 6 sportscars were enormously successful in sportscar racing, and before production ended in 1955, more than 100 had been built.

With help from aerodynamicist Frank Costin (his brother, Mike, worked for Chapman for a time), Chapman then set about developing and refining a series of aerodynamically-efficient two-seaters with all-enveloping bodywork, which proved to have quite extraordinarily high top speeds.

After the 8 and the 9, the Lotus 11 was launched in 1956. There were three versions – the Le Mans, with de Dion rear suspension, disk brakes and a Coventry-Climax overhead-cam engine of 1.1 or 1.5 liters, the Club had the same engine, but a live rear axle and drum brakes, while the basic Sports version was similar to the Club, but was fitted with a very humble side-valve 1172cc Ford engine.

Like previous Lotuses, the 11 had a multi-tube space frame chassis, now with the floor and transmission tunnel panels also acting as structural members to add to the overall chassis stiffness. The full-width style was distinguished by the use of fairing behind the driver's head.

These cars were extremely successful in motor racing, and dominated their capacity class. In 1957, for instance, four cars were entered for the Le Mans 24 Hour race, their target being the Index of Performance category, which French cars had won for several years. With a specially built 750cc Coventry-Climax engine in one car, the Lotuses took first and second in the Index, won the 1100cc class, and finished ninth, 13th, 14th and 16th overall.

A total of 150 Mark 11s were built, of which 64 were shipped to the United States, where some resourceful owners even used them as road cars, as well as for racing.

In 1957, three years after the Mark 6 had passed its best, the Seven was launched, as a development of that car, different in almost every detail. The Seven's chassis, in fact, used smaller diameter tubing, and was a development of that used in the Mark 11, but had coil spring independent front suspension borrowed from the Formula 2 Lotus single-seater of the period.

The live rear axle was suspended by combined coil spring/damper units, and located to the chassis by twin trailing arms,

Below: 1956 Lotus XI pursued by two Lotus VII, Silverstone. Colin Chapman's skill at finding engineering solutions to racing problems was already becoming a legend.
Right: 1957 Lotus XI Gullwing GT.

Right: 1953 1172cc Lotus VI. Like all Lotus cars of the time, it was built in a North London lock-up garage.

and a diagonal member. The car had drum brakes all round and, at first, a Ford 100E side-valve 1172cc engine and a three-speed gearbox. The shape of the Seven was very much like that of the 6, with square rig lines, and separate front wings, but the nose tapered a little more, and there was even space for a little luggage behind the seats. The top of the scuttle stood just 27.5 inches from the ground, and the car weighed a mere 725lb dry.

The Super Seven, which followed, was fitted with the Coventry-Climax 1100cc engine, and wire spoke wheels. Then, in June 1960, the Series 2 version appeared, in

which the chassis frame layout was simplified, there were 13in wheels instead of the 15in wheels of earlier examples, a fiberglass nose cone and front wings, and back axle location by an A-bracket. The basic car still had a Ford engine, for £587, but there was also the option of a BMC A-Series unit, for £611. In 1961, prices were cut by a full £100, and without an engine/transmission unit the Seven could be bought for a mere £399.

The most important Lotus road car so far, however, had been launched some time earlier, at the London Motor Show of 1957, this being the Elite. Not only did this car have stunning looks, as a two-seater fixed-head coupe, but it was technically unique by having monocoque construction entirely of fiberglass.

The Elite's engine was another variation on the Coventry-Climax theme, being a 75hp/1216cc unit with a single SU carburetor (though a twin-carb/85hp version was also on offer). There were all-round disk brakes and independent suspension, allied to rack and pinion steering, and center-lock wire wheels; the suspension, in both cases, was derived from that chosen for the 1957 Lotus F2 single-seater.

The Elite was an enigma. Although it was undeniably advanced, and a superbly fast car, it suffered from many detail faults, and unreliability problems. Many Elites were sold in partly-assembled 'kit form,' to be finished off by their new owners, this being part of the problem. Nevertheless, when 'on form' an Elite was a joy to use, for it had impeccable road manners, precise steering, and a splendid close ratio gearbox. Its top speed of 104mph, and 0-60mph acceleration in under 10 seconds, was quite remarkable for such a small-engined car.

The Elite also had a distinguished racing career, and in the early days, in 1958 and 1959 it allowed drivers, such as the young Jim Clark, to enhance their growing reputations. At the Nurburgring, on one occasion, an Elite beat sixteen Alfa Romeos to win the 1300cc GT class. At Le Mans, an Elite won the 1500cc class, and finished eighth overall.

In 1962, however, the Elan was announced, to take over from the Elite. This was an entirely different type of chassis, designed specifically with series production in mind, for it could be stripped out, and repaired, more easily. Colin Chapman was persuaded of the advantages of having a separate chassis, particularly in an open car. Hence the simple, but very rigid, folded steel backbone frame was conceived, and this type of construction has been used on all other Lotus road cars so far put on sale. A twin-overhead-cam conversion of the four-cylinder Ford engine, first as a 1499cc, then as a 1558cc unit, was

Left: 1972 Lotus Elan Sprint convertible.
Right: 1974 Lotus Elite.

designed by Harry Mundy. This unit was later used in the Lotus Cortina, and several other Lotus models.

Independent front suspension was by upper and lower wishbones, with a combined coil spring/damper unit, all allied to rack and pinion steering. At the rear there was what was called Chapman strut, but was really MacPherson strut, suspension, fixed to an alloy hub casting, and with a wide-based lower wishbone.

Disk brakes were fitted all round, and since those at the rear were outboard, and there were rubber drive shaft 'doughnuts' in the system, this soon gave the Elan a reputation for transmission wind-up, and the 'doughnuts' always had a limited life. The body shell was in fiberglass, but there were only two large moldings, which compared very well with the Elite, where there had been many smaller moldings.

The Elan was a great success, because of its fine combination of light weight, excellent performance, and the usual Lotus cornering capabilities, quite exceptionally better than its rivals. In 1964 an improved Series S2 Elan was announced, and a year later the more civilized fixed-head coupe also joined the range. The S3 drop head soon followed, that was overtaken by the yet more civilized S4 model and in the autumn of 1970 the Elan Sprint, with the 'big valve' 126hp engine, took over.

Meanwhile, to appeal to the family man who still wanted to enjoy the style of Lotus motoring, the Elan +2 had been introduced. This basically used the Elan-style chassis, but with a 12in longer wheelbase, wider tracks, and a more roomy cockpit in which there were '+2' seats behind the front seats. The car was only offered in fixed-head coupe form, though a few private-venture convertibles were made in later years. The Plus 2S which followed was more luxurious than the original, while the final manifestation was the Plus 2S 130,

Below: 1963 Lotus Elite (ex Team Elite). The Elite which was so successful as a race car that it became the first series production Lotus and established the still-extant link with fiberglass as a body material.

Above: Lotus XI (foreground) at Brands Hatch.

which had the 126hp engine, and the option of a five-speed gearbox.

The mid-engined Europa, first seen at the end of 1966, was at one time considered as a replacement for the Seven. The thinking was to produce an advanced, yet simple, low cost sportscar suitable for the European market. Accordingly, it was originally fitted with a Renault 16 engine and transmission, somewhat modified because that Renault was a front-engine/front-drive car.

With a similar type of backbone chassis frame as the Elan, and with supple independent suspension, the Europa had superb roadholding, as expected, but was lacking in refinement and outright performance at first. Improvements came in 1969, when the car became S2, and the car at last became fast enough in 1971 when the 105hp Lotus twin-cam engine was fitted, and the car was rechristened Europa Twin-Cam. Just a year later the Twin-Cam gave way to the Europa Special which had the big-valve/126hp engine, and a five-speed gearbox; this gave it a maximum speed of more than 120mph, and 0-60mph acceleration in 6.6 seconds. Production finally ceased in 1975, after 9230 examples had been produced.

The new generation of Lotus cars for the mid-1970s – the new generation Elites, and the mid-engined Esprits – were much larger, heavier, and more sophisticated machines. Lotus were aiming at very different customers with these cars, none of

JNG 803L

Above left: 1982 Lotus Esprit.
Below left: 1973 Lotus Europa.
Above: 1963 Caterham Super Seven Sprint produced after Lotus stopped their own production and sold the rights.

which were available in kit form. Here were cars for the well-to-do young trendy, with even more advanced technical details.

Central to the new range was the brand-new 16-valve four-cylinder 2-liter twin overhead camshaft engine, which is still the company's mainstay in the mid-1980s, and it was only appalling luck which saw Lotus launch these cars as the world of motoring fell into trouble immediately after the Middle East war and energy crisis of 1973/74. Without the usual combination of remarkable roadholding, tremendous safety qualities, distinctive styling, and excellent fuel efficiency, the company might not have survived.

Of the front-engined cars, the Elite was a close-coupled four-seater with a hatchback cabin, but a wedge-nose style, while the Eclat which followed it was a conventional 2+2 coupe on the same chassis, with the front end styling.

The Esprit, however, was a much more striking mid-engined two-seater coupe, based on a Giugiaro styling exercise, first seen on the basis of a Europa chassis in 1972. It was one of the most stylish of all the 'wedges.' Like the front-engined cars, the production Esprit, which went on sale in 1976, had a steel backbone chassis, with all-independent suspension and four-wheel disk brakes. The engine was behind the seats, and drove through a modified Citroen five-speed transaxle. The body, of course, was a fiberglass composite shell.

The result was a car very quick in a straight line, with a very high level of cornering power, and a real pleasure to drive fast. As with so many previous Lotuses, however, it was lacking in detail development and reliability, and it was not until the S2 and later the S3 came along that the concept was properly produced.

The next addition to the range, the Lotus Esprit Turbo, took the company straight into the 'Supercar' league, for the 2.2-liter engine was turbocharged, and produced no less than 210hp. However, this was much more than a tuned-up version of the old, for the chassis frame was also re-engineered, and there was comprehensive re-touching of the styling, to complement the amazing performance.

Lotus quoted a top speed of just more than 150mph for the Esprit Turbo, together with the ability to reach 60mph from rest in 5.5 seconds. This sort of performance, linked to the usual matchless roadholding standards, and surprisingly frugal fuel consumption, enabled the Esprit Turbo to stand comparison with any other exotic machine.

Maserati

The Maserati company of Italy was formed in the mid-1920s, by Alfieri Maserati, who was one of six brothers. For his marque badge, he chose Neptune's trident, the ancient symbol of Bologna, in which the works was set up.

Until 1939, most Maseratis were single-seater racing cars, and even the few two-seater sportscars were also intended for racing. The first was the 8C-1100 of 1929, which had an 8ft 2in wheelbase, front and rear tracks of 4ft 5in, and 4ft 5.75in respectively. Complete with a supercharged 1087cc twin-overhead camshaft engine, which produced 100hp at 5500rpm, it weighed 1768lb, and had a top speed of more than 80mph.

The 8C-1100 was successful in minor Italian races, but even in the same year a more powerful car, with a 120hp, 1492cc engine, was produced. Both types were endowed with square-rig bodies, which had raked back radiators, and had their fuel tanks housed in a tapering tail.

In 1930, the 4CTR-1100 appeared. This had another twin-overhead-cam engine, of 1088cc, fitted with a Rootes-type supercharger blowing through a Weber carburetor, and producing 105hp at 5500rpm. With a reduced wheelbase this car weighed just 1360lb.

Sadly for the family, Alfieri Maserati died in 1932 (though the company's title continued to bear his name), and the company continued to be run by two of the other brothers – Ettore and Ernesto.

The first semi-production Maserati road car was the A6/1500 model, announced immediately after World War II, in 1947, and built for four years. The chassis had a simple ladder-type layout, including three inch tubular side-members, with cruciform and lateral cross-bracings. Independent front suspension was by coil springs and wishbones, and the rigid rear axle was suspended on coil springs, with radius arm location.

The engine was a single overhead cam six-cylinder unit of 1488cc, and though not supercharged (it only produced 65hp at 4700rpm), it was a direct descendant of the racing unit of 1936. Most A6s were fitted with an attractive body by Farina, and their top speed was around 95mph.

The direct development of the A6 was the A6G/2000 range, built from 1951 to 1957, the first variety having a single-cam 1954cc engine, and the second (from 1954) having twin-cam 150hp 1985cc units, which were nothing less than de-tuned examples of the Formula 2 design of the day. Rear suspension on the later cars was modified, to be by cantilever leaf springs and radius arms, and top speed was of the order of 118mph, with a whole variety of body styles.

By this time the Orsi group had owned Maserati for more than a decade, and the Maserati brothers finally left to set up their own business (Osca) once again. In the 1950s Maserati built a series of successful

Right: 1956 Maserati A6 CGS Sports 2-liter.
Below right: Maserati 300S from 1957, the only year Maserati won the Constructor's World Championship.

Right: 1949 Maserati A6.
Below: 1959 Maserati 3500GT, the company's first series production car, followed Fangio's Maserati-borne World Championship.

27 ELA

8
BELL &
COLVILL LTD
8

Formula 2, Formula 1 and racing sportscars. The 300S racing sportscar was closely based on the F1 design, with the same type of twin overhead camshaft six-cylinder engine. This developed 260hp, and gave the car a top speed of up to 165mph. The next development consisted of fitting a new four-cam 4.5-liter V8 engine into the same basic chassis. The result was the brutal 450S, which had more than 400hp available, colossal performance, but only limited success in competition.

The first Maserati road car to be built in considerable numbers was the 3500GT. This was launched in 1957, but not put on sale until 1958, and – as the name implies – had a 3.5-liter twim-cam six-cylinder engine, developed from that used in the racing sportscars. With an 8.5:1 compres-

Left: 1971 Maserati Mistrale.
Below: 1979 Maserati Merak featured 3-liter alloy V6 powerplant.

sion ratio, and three twin-choke Weber carburetors, it produced 230hp at 5500rpm, and top speed was 144mph.

The chassis was a complex affair, partly tubular, partly stiffened by pressed steel members, with coil spring independent front suspension, but the rigid rear axle was suspended on half-elliptic leaf springs, with extra radius arm location. There were drum breaks at first, but Girling disks followed from 1960. Carrozzeria Touring and Allemano built the first batch of bodies, and some 100 were completed in 1959, the first complete production year. Coupe and drop head styles were both available.

In 1962, Maserati broke new ground, by offering a revised version of the car, the 3500GTI, which was fitted with Lucal fuel injection. Although this only produced another 5hp, in peak performance, it made the car more flexible, and more fuel efficient.

An even more exciting Maserati road car was first shown at Turin in 1959, this being the very limited-production (32 produced) 5000GT. Based on the same chassis and suspension layout as the 3500GT, this car had a 4935cc version of the racing V8, detuned a little for road use, but still producing 330hp at 5700rpm. It was very expensive, and a real rich man's toy.

A revised six-cylinder road car, the 'Sebring,' was introduced in 1963. Developed directly from the 3500GTI rolling chassis, it had a Vignale 2+2 coupe body, and was shorter, by four inches, than the previous model. Now with 235hp at 5800rpm, it could reach 137mph, and achieved 16.4 seconds for the standing-start ¼-mile.

In the same year, however, Maserati also launched their stylish Mistrale, still using 3500GTI/Sebring mechanical components, but with a neat fixed head, or convertible, style by Frua. Initially the car had a 3.5-liter engine, but later there was a 3692cc/245hp six, and eventually a 4012cc/255hp option, this being the final stretch of the long-serving twin-cam unit, descended from the early 1950s single-seater racing unit. The last mentioned engine gave the Mistrale a 155mph maximum speed.

Next came the Ghibli, a startlingly beautiful two-seater fastback coupe, styled by Giugiaro when he was working at Ghia. Introduced at the 1966 Turin show of 1966, this was fitted with the 4719cc/330hp, or the 4930cc/355hp four-cam V8 engine, a much changed and productionized version of that fund in the racing sportscars, and the 5000GT, and even the smaller unit was claimed to provide the Ghibli with a top speed of 174mph, which was probably over-optimistic.

Chassis design was basically that of the big, four-door saloon, Quattroporte, or its shorter-wheelbase derivative, the Mexico, and still featured the use of a rigid rear axle with half-elliptic leaf springs. Nevertheless, the Ghibli was a front-engined Supercar in every way, and the clientele loved it – 1274 were built before production ended in 1973.

Perhaps the Indy of 1969-75 was not strictly a sportscar, for it had 2+2 seating, in a sleek fixed head style by Vignale, but it was special for being the first-ever Maserati to have unit construction body/chassis engineering. Although it looked somewhat like the Ghibli, the two cars were structurally very different. Indy models were sold with 4136cc/260hp, or 4719cc/330hp V8 engines, and top speeds were up to 156mph.

In 1969 the company, perenially in finan-

Above: 1980 Maserati Kyalami was based on the four-seat Longchamps after De Tomaso rescued the company from financial collapse.

Below: 1974 Maserati 3500 GTP. After a dramatic appearance, with 5 GP wins, in 1930, Maserati were by now hovering on the edge of financial disaster.

cial difficulties, were taken over by Citroen of France, and the first true 'Citroen-controlled' Maserati was the mid-engined Bora coupe of 1971. In the evolving fashion of the day, this was a mid-engined machine, with the 4.7-liter V8 longitudinally mounted. Fitted with four twin-choke Weber carburetors, it produced 310hp at 6000rpm, and drove the rear wheels through a five-speed ZF transaxle. There was coil spring independent suspension and disk brakes all round, which gave good roadholding, but rather a firm ride.

Styling was by Giugiaro, and engineering was led by Giulio Alfieri, the 'father' of so many famous Maserati models. It had a steel unit construction body/chassis structure, and there was Citroen influence in the hydraulic 'plumbing.' It was, too, a heavy machine (weighing almost 4500lb with two passengers, their luggage, and a full 20 gallon petrol tank). On the other hand, it had a top speed of 162mph, and even though cars like the Jaguar E-Type could out-sprint it to 120mph, it was still a success.

Developed from the Bora, and first shown at the 1972 Paris salon, the Merak shared a virtually identical chassis platform, and style, but was powered by the 2965cc 90-degree four-cam V6 engine which Maserati had already developed for the Citroen SM Coupe. In the Merak it had three twin-choke Weber carburetors, and produced 190hp at 6000rpm. Because it was a shorter unit, there was more space in the cabin, so Maserati squeezed in two tiny seats for small children. The Merak was perhaps 200lb lighter than the Bora, and not nearly as fast. Its top speed was 135mph (early model), and it was nearly two seconds slower in the spring to 60mph from rest than the Bora.

A later development was the Merak SS, which had a more powerful, 220hp, engine, and a top speed of more than 150mph. Air conditioning was now a standard fitting, though the car still lacked interior space – so much so that the UK importers described it as a '2+1' model.

Citroen pulled out of Maserati in 1975, having themselves been taken over by Peugeot, and dropped the Citroen SM model, which left Maserati high and dry. Liquidation followed, but the concern was rescued by Alejandro de Tomaso, who went on to merge the company with de Tomaso, and to produce a re-engined version of his Longchamps, V8 powered Maserati, which he called Kyalami.

In passing, one should also mention the last of front-engined Maserati 2+2 coupes, which was the Bertone-styled Khamsin, also with a unit construction shell, like the Indy, but very different under the skin. As with all this breed, it used a choice of the V8 engines, and was sumptuously equipped, but also had independent rear suspension. Introduced in 1972, it was built until the early 1980s.

Left: 4.7-liter mid-mounted quad-cam V8 meant the Maserati Bora delivered 310hp, 160mph.

Mazda RX-7

By 1980, the Mazda RX-7 was the only car in mass-production to use Felix Wankel's rotary engine principle.

The Toyo Kogyo Motor Company of Japan bought a licence to develop and manufacture the rotary engine, to enhance the technical standing and staid image of their company. Starting with the R100 coupe of the late 1960s, they went on to apply twin-rotor Wankel engines to a whole range of cars in the 1970s, but their sales were hit hard in the aftermath of the 1973/74 energy crisis.

Although this eventually killed off their current rotary-engined models, as it did to every other manufacturer of rotaries in the world, Mazda persevered with the concept. It says much for the technical skill of this relatively small company, and their suppliers, that they have overcome all the technical problems, produced an utterly reliable and profitable unit, and made it the only power unit on offer in the RX-7 sports coupe.

The RX-7 was announced in 1978, and even though its engine had comparatively poor low speed torque, and mediocre fuel consumption, this was considered acceptable. The fast, attractive, and low-drag 2+2-seater soon became a cult car in the United States. Its simple and uncluttered lines remained unchanged for the first seven years of its life.

The RX-7 was built on a conventional pressed steel body/chassis monocoque, with a front-mounted twin-rotor Wankel-type engine, and rear wheel drive. The style used the familiar wedge-nose, with concealed headlamps, and there was a versatile glass hatchback. Independent front suspension was by MacPherson strut, and there was a well-located live rear axle.

The RX-7 became 'Mark 2' in 1981, when a five-speed gearbox was standardized, the power raised, and the seating and bumpers also improved. It was further improved for 1984, though the basic sheet-metal style was still not altered. Because rotary engines have excellent anti-knock characteristics, this makes them ideal for turbocharging, and this has also been an optional feature of the more recent derivatives.

Affectionately known in the United States as the 'Hummer,' the sound of the racing version of the RX-7, on open exhausts, has to be heard to be believed, and the race-tuned cars have very impressive track performances. Numerous victories have been chalked up in American IMSA events, and an RX-7 also (controversially) won the British Group 1 touring car championships in 1980. BF Goodrich also sponsor a Lola-based mid-engined racing sportscar in the USA, which uses an RX-7 engine.

Right: Rotary-engined 1979 Mazda RX-7.
Below: 1982 Mazda RX-7 Elford Turbo. Despite its modern preponderance elsewhere, there is still no factory turbo.

FKJ 418V

Mercedes-Benz

Over the years, comparatively few German manufacturers have built sportscars, but the quality has always been high. The earliest, and arguably the most distinguished, is Mercedes-Benz. Strictly speaking, the first of the line was a Mercedes, built prior to the merger with Benz in 1926. This was the 28/95 model, announced in 1921, which had a 7250cc six-cylinder engine and, unusually for the period, four-wheel brakes. Early versions were unsupercharged, but a 'blower' was later added, and Max Sailer used one of these cars to finish second in the Targa Florio.

The first of a well-known series of sportscars was the 33/180, or 'K' model Mercedes-Benz, notable for being the very first design produced by the concern's new technical director, Dr Ferdinand Porsche, who would later give his name to the famous Porsche sportscar. The 'K' had a 6.2-liter six-cylinder engine, with a single overhead camshaft valve gear layout, and was equipped with a supercharger, which was only brought into use by a mechanical linkage after the driver fully depressed the throttle pedal. This system, when in use, also differed from most other layouts, by pumping air into the carburetor, on its way to the engine, rather than using the supercharger to suck air/fuel mixture from the carburetor, and hence impelling it into the engine. This worked well enough at high engine speeds, but at low engine rpm it consumed a great deal of power. At the time the K-model was claimed to be the first-ever standard road car to have a 100mph top speed. While it was a very impressive car to look at, it suffered from having the most appalling brakes, and poor roadholding.

In 1927 an improved version of the K was introduced, this being the 36/220, or S-model. It had a larger, 6.8-liter, engine, which developed 120hp at 3000rpm without the supercharger in use, but a mighty 180hp when the supercharger was clutched into operation. This car had a claimed top speed of 110mph but, more importantly, its handling was much improved, due to a lowered center of gravity, and the brakes had also been improved. A total of 146 of these cars were produced. To prove the point, Rudolph Caracciola and Otto Merz drove 'works' racing examples, the latter winning the German GP of 1927, with other Mercedes-Benz models in second and third places.

The SS arrived in 1928, this being a lighter version of the S, in which the engine had been enlarged further, to 7020cc, and it was soon followed by the famous SSK Model, which had a shortened wheelbase, and a larger engine supercharger, so that the peak outputs were 170hp (un-blown) or

Above: 1928-30 Mercedes-Benz SSKL featured a 7.1-liter supercharged engine which developed some 225hp thanks to Ferdinand Porsche. *Left:* 1929 Mercedes-Benz 38/250 TT.

225hp (with the supercharger working). Production totals were: 112 SS, but only 33 SSK models.

Both were successful competition cars in 'works' drivers' hands. Caracciola won the British Tourist Trophy race of 1929, at 78.26mph, possibly helped by the rainy conditions, which helped reduce tire wear (and, therefore, time spent at the pits), while in an SSK the same driver won the Irish and German Grands Prix of 1930 and 1931 respectively.

The ultimate development of this family was the SSKL model of 1931, of which only five examples were ever made, for the company's own use. That car relied on the existing engine, to which an even larger supercharger was fitted, and which was reputed to make a noise like a trumpeting elephant when engaged; the peak power was 300hp, an astounding figure for the period.

The chassis was copiously drilled, to reduce weight, this being very successful, for the dry weight was down to only 2700lb. It was, in every way, a fearsome machine, with a top speed of 147mph in certain conditions; with a specially streamlined body, one example was timed at 156mph. An SSKL won the 1931 Mille Miglia, driven single-handed by Caracciola (a noble achievement) and the same driver also won the German Grand Prix, defeating, among others, two Type 51 Bugattis.

By this time, Daimler-Benz, the parent company, had decided to re-enter Grand Prix racing, and therefore had no further need to build the large, blood-and-thunder type of sportscars. For the 1930s, therefore,

Below: 1934 Mercedes-Benz 1.5-liter Type 150.

it was a very different type of road car which upheld the company's sporting image. The 500K was a 5-liter supercharged grand tourer, introduced in 1933, beautifully made and finished, but with rather disappointing performance. It was softly sprung, and more suited to the Corniche than the Nurburgring. Its overhead valve straight eight-cylinder engine developed 110hp, or 160hp with the supercharger in use. Its replacement, in 1936, was the 540K, effectively the same chassis, and with a similar appearance and choice of bodies, but a 5.4-liter version of the engine with 115hp or 180hp. Even so, this car had a maximum speed of only 105mph.

The Daimler-Benz business was virtually destroyed in World War II, and it was not until 1952 that the first postwar sportscar was ready for the public to see. Typically, however, the new Mercedes-Benz was a real *tour de force*. Not only did it have a sleek body style, with a coupe roof and lift-up gull-wing doors, but its chassis was a mass of small diameter tubes, in what is known as 'space frame' configuration. The engine on the first racing prototype had 175hp, and was a development of the new 3-liter saloon unit, but canted well over to one side to allow a low bonnet line.

In June, Hermann and Lang drove the car to victory in the Le Mans 24 Hour race, at a record 96.67mph average, and there were other wins during the season. However, it was not until the spring of 1954 that the 300SL production car was announced, with modified styling, and a more plushly trimmed interior. The engine had been further developed, had direct fuel injection, and produced 215hp from 2996cc, this

Above: 1952 Mercedes-Benz 300 SL, with fuel-injected 3-liter inline six, delivered 215hp and 165mph.
Left: 1955 Mercedes-Benz 300 SL.

giving the car a top speed of up to 150mph, depending on the gearing used. It was a formidable car, only let down by the high-pivot swing axle rear suspension, which could produce vicious oversteer in hard cornering conditions.

The coupe gave way to an open roadster from 1957, with conventional front-hinged doors, and an optional hardtop was available. More important still, this derivative had much more effective low-pivot swing axle rear suspension. By the time production ended in 1963, 3250 cars had been built.

Also in this period, there was the 190SL model, which lacked the glamor of the larger car, and was actualy based on the underpan, engine, and suspensions of the 180 saloon of the day. Because it was too heavy, and too costly, it lacked the attraction and the performance of British mass-produced sportscars of the day, yet sold well because of its high-quality trim and finish.

In this period, Mercedes-Benz also produced a ferocious racing sportscar, used in only the 1955 season, called the 300SLR. This was based on the general layout of the

Below: 1957 Mercedes-Benz 300 SL Roadster, still a powerful and expensive supercar, soon to be replaced by the more conventional 230.

Above: Mercedes 350SL 3.5-liter V8 Roadster took Mercedes back into the supercar bracket on price if nothing else.
Right: 1963 Mercedes 230 SL.

then-dominant W196 Formula 1 single-seater, but naturally had a different space frame chassis to allow two seats, and an all-enveloping body style.

Its 3-liter straight-eight engine had desmodromic (positive opening and closing) valve gear, and produced 300hp, which gave the car a top speed of 180mph. Perhaps its only real weakness was that drum brakes were used, at a time when Jaguar's D-Type was revolutionizing motor racing with its disks. To give extra braking at Le Mans, Mercedes-Benz provided a massive hinge-up air brake across the tail, behind the cockpit, which was not only effective from high-speeds, but when in operation it put considerable down force on the rear tires, and increased the overall cornering power. A total of ten cars were built, of which two (unraced) were almost 300SL 'gullwing' lookalikes. In one season, the team cars entered six events, won five, and were withdrawn from Le Mans when one of them plunged into the crowd opposite the pits, killing more than 80 people.

In 1963, both the 190SL and the 300SL were replaced by a single car – the 230SL – which owed more in layout and philosophy to the small car, in that it had a conventional pressed steel body/chassis hull, and suspension and engine components lifted from the latest 220SE saloon model. The engine, at first, was a 170hp/2306cc overhead-cam six-cylinder unit, with indirect fuel injection, there was power-assisted steering (unusual for this period), and the usual type of low-pivot rear suspension.

To many people, this model is best remembered by the unusual 'pagoda' style of its optional hardtop, and there was some sales resistance to this feature at first. However the car's image was instantly improved when the 'works' rally driver, Eugen Bohringer, took one on the Liège-Sofia-Liège marathon rally of 1963, and won it outright. Two developments of this car were produced – the 250SL, built only in 1967, and the 280SL (with 2778cc engine) produced from 1968 to the end of production in 1971.

Then came the 350SL, to replace the 280SL, entirely different, much larger, much 'softer' still in its image, but directed at the same luxury-conscious type of clientele. The monocoque used a drive line and suspensions from current saloon car models, notably the still-to-come S-Class of 1972, and there was a single-cam-per-bank V8 engine of 3.5 liters, which produced 220hp at 5800rpm, and allowed a maxium speed of 130mph. It weighed all of 4000lb, which meant that it needed, and had, massive four-wheel disk brakes. A 4.5-liter V8 engine was soon optional (standard in the United States), and long-wheelbase, 2+2-seater, fixed-head coupes known as

350SLC or 450SLC were also made available. At the end of the 1970s, the latest light-alloy version of the V8 engines were fitted, of 3.8 and 5.0 liters, and these cars were still made in the mid-1980s.

While these cars were still under development, the company also indulged themselves with the development of an ultra-high-performance prototype, the Wankel rotary engined C111. This was a state-of-the-art mid-engined two-seater coupe, with all-independent suspension. The original C111 of 1969 had a three-rotor engine of 3600cc nominal capacity, which developed 280hp, and this gave the car a top speed of 160mph. The bodyshell was in fiberglass, and gull-wing lift-up doors were fitted. Within a year the next evolution of this design appeared, having an amazing four-rotor engine, of 4800cc (nominal) and 350hp. That car was good for 190mph. The C111 project was fascinating, and demonstrated the way that Daimler-Benz engineers can achieve almost anything to which they set their minds, but the cars were never put into production.

Right: One of several versions of the experimental Mercedes C111 which appeared during the seventies but which never made it to production.
Below: 4-rotor Wankel-engined version of the Mercedes C111.

Mercer

The Mercer Automobile Co., logically enough, was named after Mercer County, in New Jersey, USA, and the factory was actually in Trenton, close by. The marque is best known for is Type 35 Raceabout, introduced in 1911. This car was designed by Finley Robert Porter, but he soon left the company, to make his own FRP, then Porter, luxury cars. Mercer's financial backing came from Washington A Roebling, who was drowned when the Titanic sank.

Early Raceabouts looked rather similar to the Stutz Bearcats, but had a cooling radiator reminiscent of the Mercedes of the day. They had two bucket seats, an exposed cylindrical-section fuel tank behind them, and no weather protection of any sort. The engine was a 5-liter, four-cylinder Continental unit, called T-head because it had lines of side valve at each side of the bores, and this gave the car a top speed in the 70s, from 55hp. Later, in 1915, a more efficient 90hp L-head (all valves down the same side of the engine) was adopted, along with a four-speed gearbox.

Even though they sold fewer than 500 cars every year, Mercer enjoyed a fame far in excess of this modest output. This was mainly due to their extrovert appearance, and to the glamorous young people who seemed to buy, and be seen, in them.

Over the years the Raceabout was given various improvements, but changed little in appearance, except for gaining sides to its bodywork. In 1918, it was available in yellow, gray or gunmetal finish, while one could also have the Runabout model, which had windscreens and fold-down hoods. All of these cars, however, were rated at the same 22.5hp (RAC rating).

Soon after World War I, a former Packard salesman, Emlen Hare, acquired three American marques – Crane-Simplex, Locomobile, and Mercer – to weld them into a conglomerate, but this enterprise failed, and liquidation followed in 1921.

The Mercer name, however, was rescued, and henceforth used Rochester six-cylinder overhad valve engines. However, by the mid-1920s, the demand for traditional-style roadsters had virtually gone, and as the company had never successfully broken into the touring car market, it collapsed again in 1925.

There was still sufficient magic in the name for one more try, so in 1931 it was revived by the Elcar Motor Car Co. of Elkhart, Indiana, who spoke of receiving orders for 750 cars, and of 3000 cars to be produced in a year. The new car was to have a 5.3-liter straight-eight Continental engine. The new Mercer was launched with a flourish from the Hotel Montclair in New York, but Elcar was so weakened by the Depression that it could not even continue building its own cars, let alone introduce the Mercer, so the marque finally died out.

Right: 1922 Mercer Roadster.
Below: 1913 Mercer Raceabout was the arch-rival to Stutz on the road and racetrack. Rival owners even composed impolite slogans about each other: 'You have to be nuts to drive a Stutz' and 'There's nothing worser than a Mercer' were the best-known.

6E 52 84
19 CALIFORNIA 37

MG

Many a bet has been won and lost about the origin, and meaning, of the world-famous MG initials. In fact, and quite simply, they stand for Morris Garages.

In 1923 Cecil Kimber, who was currently general manager of that concern, began to produce and sell Morris Cowleys, with close-coupled four-seater bodies made for him by Carbodies of Coventry. However, when Morris Motors themselves began to offer a very similar style on the Cowley, Kimber then turned to Raworth of Oxford, who produced six two-seater tourers, followed by a small series of four-door saloons based on the Morris Oxford chassis, and it was these cars which were advertised in *The Morris Owner* magazine for March 1924 as the 'MG V-front saloon;' this was the first-ever appearance of the initials in print. Two months later the Raworth two-seater, which must have been selling very slowly, as the first had been delivered in August 1923, was being advertised as the 'MG Super Sports Morris' – so perhaps either of these two models can be considered as the very first MG.

During 1924, a small series of sporting four-seaters, based on the Morris Oxford, was built, these cars having mildly tuned engines, lowered chassis, and polished aluminum Ace wheel disks covering the 'artillery' style wheels of the standard car. William Morris, who owned Morris Motors, *and* Morris Garages (though they were still separate businesses), agreed to a limited production run of this model, which was also called the MG Super Sports.

The well-known, starkly-trimmed, two-seater, registered FC 7900, has often been claimed as the 'first MG', but it was nothing of the sort. It was actually a one-off special for Cecil Kimber, completed in early 1925, with a special overhead-valve version of the Morris's 'Hotchkiss' engine design – and no MG production ever had overhead valves until the arrival of the 18/80 model of 1928.

By 1926 demand for these cars had grown so much that larger premises were needed. Production moved to Bainton Road, in North Oxford, and it was here that the first 'production' MG, the 14/40 was produced. That car had a standard Morris chassis frame,

Above right: 1937 MG TA, the opening model in the collectible and well-renowned T Series MGs.
Far right: 1945-8 MG TC produced under Nuffield's 'Export or die' regime.
Below right: 1933 MG K3 Magnette.
Below: 1931 MG 'C' Type Montlhéry Midget.

JNY 352

but modified and harder suspension, a special back axle, and a tuned-up side-valve 1.8-liter Morris engine. Maximum speed was about 65mph, though the car could cruise at 60mph. As it weighed under 2200lb, the petrol consumption was at least 30mpg.

Once again MG grew out of its premises, so a move was made to Edmund Road, not far from the big Morris factory in Cowley, but even this facility did not last long, and in 1929 they moved again, to a disused leather factory at Abingdon, just a few miles south of Oxford itself.

In the autumn of 1928, however, MG had introduced two new models. The first was the 18/80 model, entirely different from anything that had gone before. Only the engine

Right: 1955 MG TF was greeted with horror by MG enthusiasts at the time thanks to steel wheels and inboard headlamps.
Below: 1962 MGA 1600 Mk II.

remained Morris, and even this was a 2468cc six-cylinder overhead-cam design produced at Hotchkiss with MG in mind! It was also intended for use in Morris Light Six/Six/Isis models of the late 1920s, but all suffered from poor chassis and roadholding. For its MG application this engine was tuned up, and fitted with twin SU carburetors; on the chassis, Rudge-Whitworth center-lock wire wheels were used for the first time on an MG. A total of 736 of these cars, Mks I and II, were built up until 1933.

Also in 1928 there was the new model which founded the large-scale production success of the company, the M-Type Midget. The chassis and running gear was almost exactly that of the new Morris Minor model, which had a single-overhead-cam (Wolseley-designed) 847cc engine, and the original cars had a wooden-framed body shell, fabric covered, with what is now the characteristic MG radiator shell, and cycle-type wings. Even though it only produced 20hp, the engine only had to propel an 1100lb car, so it provided brisk performance.

Developments included the use of a 746cc engine for competition use (in the 750cc class) against the Austin Seven, and for the Montlhéry derivative this could be supplied in normal, or Powerplus supercharged form. The blown cars produced up to 60hp, and had 90mph top speeds. Later cars, of all types, had different body styles, some with pointed tails, and light metal panelling.

In 1932 the M-Type, or 8/33 as it had also been called, was dropped in favor of the new J2 Midget, which was the epitome of a style so typical of all British sportscars in the next decade, for it had a slim shell, cutaway doors, a slab fuel tank, and a spare wheel mounted out on the tail. There was a rare J1 saloon version, and two special supercharged racing versions known as J3 and J4. However from early 1934 the J2 gave way to the more sophisticated P-Type (later known as PA) Midget, which had more flowing styling, and other creature comforts. Like all other MGs of the period, though, it had an overhead camshaft engine, rock hard leaf spring suspension, and a 'crash' (non-synchromesh) gearbox.

The last in this series of cars was the PB, introduced in 1935, which had an enlarged (939cc) engine producing 43hp at 5500rpm, and we now know that it was to be the last of this line. All the engines came from Wolseley, with special modifications, though MG sometimes went to great lengths to disguise this fact.

Racing Midgets gained many competition successes, for Kimber was very keen on this activity as a way of publicizing and improving the breed. In 1930, for instance, the MG team took the team prize in the JCC Double-twelve (hour) race at Brooklands, while in the same race in 1931 they took the first six places in their class *and* won the race outright on handicap. A Midget also won the Irish Grand Prix, and the Tourist Trophy race in 1931, and in 1932 cars took the team prize and class victory, and outright victory for R T Horton's special-bodied car at 96.29mph. The Midget, in tuned form, was also the first 750cc engined car to exceed 100mph, and to cover 100 miles in an hour.

In parallel with the Midget, too, MG also manufactured a small six-cylinder-engined range. First, in 1931, was the F-Type Magna, which had a 37hp version of the Wolseley Hornet engine (Morris had owned Wolseley since 1927), of 1271cc, with overhead-camshaft valve gear. It used a four-seater sports body of low build, but was only capable of about 70mph. In due course, in 1933, it was supplanted by the L-Type Magna, which had a more specialized engine, with a cross-flow cylinder head, and carburetors now on the right.

The K-Type Magnette series arrived in 1932, and the K3 sports car of 1933 covered itself with glory in the racing world. These cars used a 1087cc short-stroke engine, but other Magnettes used the original 1271cc size. Early less-specialized versions were K1s and K2s, being sold with open and closed body styles, and the N-Type Magnettes arrived to take over, with 57hp instead of 41 or 48hp, in 1934.

Above: 1971 MG Midget. Despite a number of 'hot' saloons given the MG badge since, this was the last sportscar to bear the hallowed Octagon.

The K3's greatest claim to fame was in an outright victory in the 1933 Tourist Trophy race on the Ards circuit in Northern Ireland, when driven by the great Tazio Nuvolari. It had a powerful supercharged version of the 1087cc engine, and was fitted with a preselector transmission. In stripped out form a K3 was even competitive for single-seater racing car events, for Whitney Straight won the Coppa Acerbo Junior event of 1933, against single-seater Maseratis on their home ground.

Only 33 K3s were built, and these were followed by ten single-seater R-Type models, which had a backbone chassis frame, all-independent suspension, and supercharged 746cc engines.

In 1935 there were corporate changes at Morris Motors, and one result was that both Wolseley and MG were taken into the direct control of that firm. The Abingdon design department was closed down, and for mid-1936 the Cowley design offices produced the new TA Midget, which had an overhead valve 1292cc engine tuned from that of the new Wolseley 10/40 model and (after the first few cars had been built) a gearbox with synchromesh on the upper ratios. The TA was a somewhat 'softer' car, easier to keep in tune than the PB which it replaced at Abingdon, though the engine did not rev as readily.

For 1939 only, the TA became TB, with the aid of a new type of short-stroke 1250cc engine (basically that of the new Morris 10 'M'), and would have gone on to sell well if the outbreak of war had not closed down private car production. Almost all TAs and TBs were traditionally-styled open two seaters, though there was also the option of a coachbuilt Tickford model, with wind-up door glass, and a fold-down hood.

After World War II, MG restarted production of the TC sportscar in 1945, this being a very lightly modified TB. Although produced only in right-hand-drive, it introduced MG to the USA market. 10,000 TCs were produced up to the end of 1949, and many were used in competitions.

The Americans, however, wanted more comfort, if not more modern looks, and this led to the introduction of the TD of 1950. This had a new box-section chassis frame, with coil spring independent front suspension, rack and pinion steering, and a hypoid-bevel rear axle, plus a more spacious, but still traditionally-styled, body. Left hand drive, but no wire-spoke wheels, was available, and sales boomed. In 1952, the TD's peak year, 90 per cent of cars were exported, and a total of 29,664 were produced in four years.

By this time MG, already a part of the Nuffield organization, had been swallowed up by the British Motor Corporation concern. Although a replacement for the TD was designed (this would become the MGA in later years), it was frozen out by BMC in favor of the new Austin-Healey 100, and in order to give MG breathing space a face-lifted car, known as the TF, was hurriedly produced for the autumn of 1953.

This was a disappointment to its public, as the chassis was still the same as the TD, and it could still only achieve 80mph. Even with the larger, 1466cc engine introduced in mid-1954, it was still no match for cars like the Triumph TR2, and it faded out in the spring of 1955.

The new MGA, designed in 1952, was finally revealed in 1955, and was an instant success. Equipped with a very strong separate chassis, and the TF's front suspension, it had a tuned-up BMC 1489cc push-

rod overhead valve engine, and a beautiful all-enveloping body style, which helped it to a top speed of nearly 100mph. A comfortable bubble-top fixed head coupe was announced a year later, and a detachable hardtop was always available.

Original cars were underpowered against the 1991cc TR2, or the 2.6-liter Austin-Healeys, but the 1588cc model of 1959, and the 1622cc Mk II of 1961, helped improve the performance. The last of all were built in 1962, much the most successful MG Model so far, for 101,081 had been produced, of which only 5815 had originally been delivered in the UK.

The company originally planned an extensive racing program for this model, and introduced it as a prototype at the 1955 Le Mans race, but this was axed by management after the fatal crashes (not involving MG) at Le Mans, and the Dundrod race which followed. However, MGAs were used in European rallying, and Nancy Mitchell won the European Ladies' Championship twice, in 1956 and 1957.

The MGA Twin-Cam was a limited production model (2111 built), with a specially-developed twin-overhead camshaft engine of 108hp, and four-wheel disk brakes, but unfortunately it acquired an unreliability reputation which proved impossible to live down.

The MGA's successor, introduced in 1962, was the MGB, and this featured a unit-construction body/chassis design, though the same front suspension, and basic engine/drive line remained as before. The engine had been enlarged to 1798cc, and produced 95hp, so top speed was comfortably over 100mph. At first only a two-seater open sports version was available, but the very elegant MGB GT hatchback derivative followed in 1965.

The MGB was an amazingly long-lived design, for the last was produced in 1980, and a grand total of 512,880 were eventually built. It was unfortunate that the design was improved very little during that time, for in the last few years it lagged well behind its competitors.

Two limited-production models were developed from the MGB. The first was the MGC of 1967/69, which used a six-cylinder overhead valve 145hp engine developed from that used in the Austin 3-liter saloon, along with torsion bar front suspension, but this always had too much understeer, and was dropped after 8999 examples had been made.

The MGB GT V8 of 1973/76 was a much better car, even though it was too expensive, and only 2591 examples were sold. Based on the GT hatchback body, it was equipped with the 137hp version of the Rover (ex-GM) light-alloy V8 engine. It was a very fast car, which could exceed 125mph, with silky refinement, but perhaps it was produced five years too late, for MG and Rover had been linked (in British Leyland) since 1968.

The four-cylinder MGB was a very successful long-distance racing sportscar, and appeared three times at Le Mans, winning its class in 1963 at 92mph, and improving on that to 98.26mph in 1965.

Although one must not forget the long-running Midget of 1961/79 (226,526 built), it was really no more than a badge-engineered Austin-Healey Sprite. From late 1974, a Triumph Spitfire 1491cc/65hp engine was used, which did not endear it to the die-hards, but this helped raise the top speed to a genuine 100mph, the first time ever on a Midget.

Under British Leyland management, MG was persistently neglected in favor of Triumph, so no new sportscars came along to replace the Midget and the MGB after 1980. The MG badge has been resurrected, on Metro, Maestro and Montego models, but the Abingdon spirit has now been lost. Nevertheless, the Metro Turbo and Montego Turbo are the first turbocharged motor cars ever to carry this illustrious badge.

Above: 1965 MGC convertible produced yet more controversy, though it offered the power boost the MGB needed.

Left: 1965 MGB raced at Le Mans by top international rally-driver Paddy Hopkirk.
Below: 1965 MGB convertible.

Morgan

A well-known motoring historian once quipped that Morgan are alive, and well, and living in the 1930s, which really does sum up their design philosophy rather well. Nothing on the Morgan motor car has ever been changed for the sake of change. In all but price, performance, and visual detail, the Morgan of today is much the same sort of vintage-style sportscar which was first produced in the 1930s.

The company was founded by H F S Morgan, who produced his first crude, but speedy, tricycle, in 1910. These three-wheelers – two front wheels, one driven rear wheel – were persistently developed, and gained great popularity, both for their economy, and their sporting pretensions.

Below: 1933 Morgan. The famed JAP engines were replaced by Ford sidevalve units.

Some had specially-built Anzani units, but a majority had JAP units, either air-cooled or water-cooled, mainly vee-twins of around 1000cc. The Super Sports model, with Matchless MX4 engine, had more than 40hp and this, combined with minimal weight, helped produce a very lively performance. Fully-tuned 'Grand Prix' Morgans were capable of 115mph.

The later type of three-wheeler, introduced in the early 1930s, used a side-valve Ford engine, and a more solid chassis frame, and in this form continued to be built until the early 1950s.

In 1935, 'H.F.S.' decided that Morgan would have to manufacture a four-wheeler car for his company to stay in business, and the result was the original 4/4, launched at the end of 1935, which used a 1.1-liter Coventry-Climax engine, with overhead inlet and side exhaust valves, producing 34hp, and this endowed the car with a top speed of around 75mph.

There was a simple ladder-style chassis, underslung at the rear, and with Z-section side members, a profile quite unique to Morgan, and retained ever since. At the front there was the familiar Morgan independent suspension, consisting of vertical sliding pillars, and coil springs – a layout found on the first Morgan of all, and never changed since then. There was a Meadows four-speed gearbox, this being separately mounted from the engine, but connected to it by a long input shaft encased in an alloy tube.

The 4/4 was built until 1950, though a specially-produced overhead-valve Standard engine of 1267cc was made available from 1939, and standardized from 1945. This had evolved from the Standard Flying Ten unit, which was a side-valve unit. Also from 1939, a Moss gearbox took over from the Meadows item, but still separate from the engine.

It was then decided to take the big step of producing a much bigger-engined, and faster car, the result being the Plus 4, which had the Standard Vanguard's 2088cc four-

Above: 1926 Morgan Super Sports. Despite unusual 'tricycle' arrangement, they were popular club racers for decades.

cylinder engine, which produced 68hp at 4300rpm. For the Plus 4 the wheelbase was lengthened, and some detail changes were made to the suspension, but there were no major modifications. Because the Plus 4 weighed only 1800lb, it accelerated much faster than the previous model, though the poor aerodynamics restricted maximum speed.

As ever, the gearbox was separately mounted from the engine, the ride was hard and uncomfortable, and the roadholding was splendid. As ever, too, there was a choice of body styles on the same chassis – two-seater open sports, four-seater open tourer, and two-seater drophead coupe.

From 1954 the Plus 4 offered the 90hp Triumph TR2 engine of 1991cc (a modified Vanguard unit, of course) as an option, and this eventually became standard. It made the car a near-100mph machine, which could reach 60mph from rest in just over 10 seconds, and – most importantly – it slipped neatly into the 2-liter competition class.

From this point on, therefore, the Plus 4 became a successful club racing/rally car in the UK, and driving test machine. H F S Morgan's son Peter soon became a star driver in rallies and long-distance trials, in one of his cars. By 1956 the Plus 4 had 100hp, for the TR's output had risen slightly, and it was a formidable, if old-fashioned looking, sportscar.

In 1953 and 1954, however, the looks were changed twice, first to provide a sloping, though flat, radiator style, then to provide the more familiar cowled style which has been with the Morgan ever since. At the rear, one of the two spare wheels was discontinued, and the remaining spare was sloped forward, and partially recessed into the tail panels. In the course of this change the headlamps were partially recessed into the panel between bonnet and wings.

In 1955, however, Morgan re-introduced the 4/4, this using the same 8ft 0in wheelbase frame as the Plus 4, but being fitted with the Ford 1172cc side-valve four cylinder engine, and its own three-speed gearbox. This was the first Morgan four-wheeler to have a gearbox integral with the engine. Five years later, as Ford changed their models, so did the 4/4 inherit the new overhead valve 997cc engine unit.

In 1963, however, Morgan introduced their least successful car of all time, the extraordinary-looking Plus Four Plus, which had a closed two-seater coupe body style, the body being all-enveloping, and built of fiberglass. The traditional Morgan customer did not fancy this at all, and it took four years to sell just 26 examples. No such flights of modernization have ever been entertained at Malvern Link since then!

During the 1960s, while the design of the Plus 4 stagnated somewhat after the front wheel disk brakes and the 2138cc Triumph TR4 engine had been finalized (plus a few highly-tuned Lawrencetune 116hp engines), the 4/4 gradually 'grew up', for the 997cc engine was replaced by a 1340cc unit, then a 1499cc unit, to Cortina GT tune, and finally, in 1968 to 1599cc, with the latest bowl-in-piston Ford 'Kent' engine. This car was by this time a 100mph machine.

From 1968, even though the 4/4 sold, and has kept on selling, it was overshadowed by the impact of the Plus 8. Caused because of the need to replace the Triumph TR4 engine, which was discontinued at Triumph in 1967, it inspired Morgan to test, and subsequently gain supplies, of the light-weight new Rover V8 engine of 3528cc.

To instal this bulky engine, it was necessary to lengthen the wheelbase by two inches, there was a rather wider body to give more under-bonnet space, and more

Below right: 1928 Morgan Super Sports.
Below left: 1935 Morgan.

robust front suspension, and rear spring mountings. The Plus 8 was only sold with the open two-seater sports body style.

At first the Plus 8 retained the separate Moss gearbox, but from the spring of 1972 this was dropped, and the cars were built with the Rover 3500S type of four-speed all-synchromesh gearbox in unit with the 151hp engine. Then, from the beginning of 1977, this box was replaced by the new five-speed transmission as fitted to the big Rover 3500 hatchback, and at the same time the latest 155hp engine was fitted.

In that form, the Plus 8 was a very fast car indeed, with a top speed of 123mph, and very rapid 0-60mph acceleration capability of about 6.5 seconds.

However, although nothing ever changes at Morgan, something *is* always changing! The 4/4 was re-engined from the beginning of 1982, when the old 'Kent' engine was dropped, and a choice of engines – 1.6-liter single-overhead-cam Escort XR3 engine, or 1.6-liter twin-overhead-cam Fiat twin-cam power – was made available, each with their appropriate integral gearboxes, and for 1985 there was also the possibility of a 2.0-liter Fiat twin-cam also being fitted.

The Plus 8, meantime, continues, but the more powerful 190hp Rover Vitesse V8 was also offered, from £13,996. The latest Morgans even had rack-and-pinion steering. Whatever next?

As ever, incidentally, there were still long waiting lists for Morgans – several years, not months.

Top left: 1977 Morgan Plus Four SS.
Left: 1976 Morgan Plus Eight.
Below: 1951 Morgan Plus Four.
Right: 1972 Morgan 4/4. Traditional coachbuilding methods still in use keep demand for this series constant.

DORMER COTTAGE
RNP 352L

Plymouth Road Runner

Road Runner was a cartoon character created by Warner Brothers, in Hollywood, and based on the real live running bird to be found in the western states of the USA. 'Road runner' was extremely fast, always eluded his deadly enemy, the Coyote, and had that inimitable communication of 'Beep beep.'

Chrysler, therefore, showed a rare flash of humor, when they named a new muscle car 'Road Runner' in 1968, badging it as a Plymouth. This car had a compact unit construction body/chassis unit, with a 383cu. in./6.3-liter V8 engine as standard equipment, or the famous 426cu. in/7.0-liter Chrysler 'Hemi' V8, of 425hp as a very popular option.

To try to keep the mighty torque of the 426 Hemi in check, the car was supplied with six-leaf half-elliptic springs to the left side, but a seven-leaf spring to the right side, yet disk brakes were not available. Road Runner was a light, 'no frills' muscle car, and was made Car of the Year by Motor Trend in 1969.

The 1970 model was also made available as the Superbird, billed as the 'Ultimate Road Runner,' having the same shovel nose, and high, full-width, rear aerofoil as the NASCAR models, and the Dodge Charger Daytona whose body shell it shared, but this lasted only one season. For 1971 the nose of the car was re-styled, with a front bumper wrapping right around the grille, and this was the last year in which the thirsty Hemi engine was available.

All pictures: 1970 Plymouth Road Runner Superbird carried aerodynamic droop-snoot and tall rear wing to give it a genuine 200mph on the NASCAR ovals.

CAUTION-FAN

ROAD RUNNER
SUPERBIRD
Plymouth

JUL
810N

Pontiac Firebird

The Firebird was introduced at the very beginning of 1967, and shared its sporty two-door close-coupled four-seater monocoque body with the Chevrolet Camaro, but was more expensive, better equipped, and fitted with its own Pontiac engines. As with the Camaro, it was a direct competitor of the Ford Mustang.

Originally offered with a choice of overhead-cam straight six, or large-capacity V8 engines (the largest being a 6.55-liter 325hp unit), the Firebird was available as a coupe, or a convertible. Because it had the traditional type of Pontiac split grille, cleverly combined with the front bumper, it looked more distinguished than its sister Camaro, and – like the GTO in the same year – it had a bonnet-mounted rev-counter as an option.

For 1968 the side window quarter windows were discarded in favor of one-piece door glass, while the 1969 models shared the Camaro's style changes behind the wheel arches. In 1969, too, the legendary Trans Am appeared for the first time, complete with a special ram air bonnet with twin scoops, louvers behind the front wheel arches, and a large rear spoiler across the tail. Top engine option was a 355hp 400cu. in./6.55-liter V8, and front disk brakes were standard.

For 1970 the Firebird, like the Camaro, was completely restyled, and many believe that a classic shape resulted. It was only built as a coupe, though the successful engine and mechanical options were all retained. Forward facing scoops were retained for the Formula 400 version, while the Trans Am got a deep front spoiler under its Endura nose, air deflectors ahead of each wheel opening, and a large rear spoiler, the last set to give a 50lb down force at high speeds. Most impressive of all, for the occupants, was the rearward-facing 'shaker' bonnet scoop, really an extension of the engine air cleaner, which moved sideways as the engine twisted on its mountings, under load.

For 1971 the larger 455cu. in./7.45-liter V8 was introduced, but the body remained fundamentally unchanged. During the 1970s the Firebird's style was gradually changed, mirroring those made to the Chevrolet Camaro, and engines became less powerful as the emission laws bit hard. There was a wrap-around rear window for 1975, and a new style nose, with rectangular headlamps, for 1977. The Trans Am was adopted as the Indianapolis Pace Car for the 1980 race, so the factory produced replicas for sale, with turbocharged V8 engines (5-liters, and 210hp), and a host of options including cast aluminum wheels.

For 1982 the Firebird was completely and dramatically restyled, once again as an upmarket version of the sensational new Chevrolet Camaro. In fact it scored over the Camaro by having a more streamlined nose, with concealed headlamps, and this type gained immortality in many countries as the star of the TV series 'Knight Rider.'

In 1985 the Trans Am gained the eight-port fuel injection system of the Corvette for its Chevrolet-built V-8 engine, which boosted the output of the 5-liter unit to 210hp. Nevertheless, the 'base' engine of the Firebird was still the humble 2.5-litre 'Iron Duke' four-cylinder unit.

Left: 1975 Pontiac Firebird 400 with 'Endura' nosecone and overall restyle introduced in 1974.
Below: 1969 Pontiac Firebird convertible.

Pontiac GTO

It was Pontiac which started the USA's 'muscle car' craze, and more particularly it was John Z DeLorean who saw that by putting the biggest Pontiac engine – the 455cu. in./7.45-liter V8 – from the full-size Grand Prix model, into the compact Tempest shell, he could create a car which had quite staggering acceleration, even if it was short of brakes, or the sort of roadholding that any European car with the same power/weight ratio might expect to have. The handling of this hybrid, in fact, was still on a par with that of any other American car of the period.

DeLorean's move was made because Pontiac, at this time in the early 1960s, was losing market share to other manufacturers, and management felt that they needed to attract the younger class of driver to the Pontiac badge. The engine transplant idea worked well, for young men were attracted to the GTO for its sheer performance.

The square-cut 1964 model was followed by a most desirable classic design for 1965, in which the two headlamps each side were stacked, one above the other. The cars which followed, in 1966 and 1967, had a more rounded 'Coke bottle' rear body lines, which were emphasized by a shaped stainless sill moulding under the doors.

Disk brakes, fortunately, became available for the front wheels of 1967 models, as an option, which at least brought the 'stop' side of performance into line with the 'go.' These were modified Corvette brakes, with four-pot calipers, and were shared with the Firebird and the Chevrolet Camaro.

For 1968 the GTO was given a less appealing Endura nosed notch-back style of body, but for 1969 and 1970 there was the extrovert 'Judge' (Here Come De Judge . . .) version, with most options as standard equipment.

1970 was the last year for GTO Production, for the newly styled Firebird, which appeared in the same year, neatly replaced the original type of muscle car. The Pontiac, however, will always be remembered as the brash model which started the short-lived muscle car trend, and boldly adopted the famous Ferrari initials of GTO, which mean Gran Turismo Omologato.

Right and below: 1969 Pontiac GTO 'The Judge' was a simple styling exercise designed to rekindle demand for the car which had started the musclecar ball rolling.

THE JUDGE

Porsche

After designing many famous cars, including a string of fast Mercedes-Benz sportscars, and the German people's car, the Volkswagen 'Beetle,' Dr Ferdinand Porsche spent the years of World War II working on military machinery, then was unfortunately imprisoned in France immediately after the war. It was during that period, however, that his son Ferry set about producing a new sportscar, initially in Austria, which would carry his name. Not surprisingly, it utilized many VW components.

The new car was given the type number '356', and had a sheet metal chassis platform, with a fabricated section at the rear to carry the engine and transmission, and the rear suspension. Initially the engine was a 1086cc version of the flat-four, air-cooled, VW unit, and it was mounted behind the line of the rear wheels. With special cylinder heads, it developed 40hp, and the car's top speed was 80mph.

The first coupe was completed in August 1948, and rather spasmodic production began at Gmund, in Austria (Porsche himself was Austrian by birth). However, from 1950 the business was moved to Stuttgart, in West Germany, where the elder Porsche had spent so many years of his life, and the car's bodies were made from steel, rather than from aluminum as in the first batch.

In April 1951 a 1.3-liter engine, producing 44hp at 4200rpm, was fitted to the cars, and only six months later a 1.5-liter version was also offered, these 356s being available in open cabriolet or closed coupe form.

In the next few years this model evolved gradually, with detailed but significant changes, rather than by radical redesign. In mid 1952 the bumpers were separated from the body, and a one piece windscreen replaced the original screen. Then a new transaxle, with synchromesh, and a higher torque capability to deal with the 1500's 70hp, was designed by Porsche, but manufactured by Getrag. Then, inspired by Porsche's East Coast USA distributor's desire to have a cheap Porsche to sell, the factory then created a bulbous roadster known as the Speedster. This had a very shallow windscreen, and a claustrophobic hood and small side screens, but because it was so light it was a very successful competition car.

In 1955 the car became the Type 356A, and boasted an optional 1.6-liter engine, and at the Frankfurt show of that year there was a new model derivative with the name 'Carrera' – one which was to crop up again and again on Porsches over the years. The original Carrera, basically, was a 356 with a very special new four-overhead-cam engine, with cam drive by shafts and bevel gears, twin sparking plugs and a roller bearing crankshaft. Initially, in 1.5-liter form, it produced 110hp at 7000rpm, but later there was a 1.6, and finally a 2.0-liter version. The last car, so equipped, was called Carrera 2, and was capable of 130mph. In due course the ultimate Carerra type was a lightweight car titled the GTL, or Abarth-Carrera, with a differently styled aluminum body by Zagato.

The 10,000 Porsche was built in 1956, a year later the 1300cc engine was dropped, and diaphragm clutches were adopted. A further year saw the Speedster replaced by the Speedster D, soon to be renamed the convertible D.

Below left: 1961 Porsche 356 (left) and 1971 Porsche 911T (right).
Above: 1955 Porsche 356 Speedster was one of America's favorite sportscars.
Right: Porsche 356 Speedster pictured competing in the 1964 Monte Carlo Rally.

The 356B derivative had a slightly changed body style, with a flatter nose but more pronounced headlamps, and this gave a hint of the 911 model, which was yet to come. In 1960 the Super 90 version of this car went on sale, having a 90hp engine (and a top speed well over 100mph), revised rear suspension, and radial ply tires plus Koni dampers, to help improve the road-holding.

The final manifestation of the type was the 356C, available for 1963, in which disk brakes were added, and where derivatives were known as 1600C and 1600SC. Production of all 356s ceased in 1965, after a grand total of 76,303 of all types had been made.

To replace the 356, but actually to run alongside it for a short period, Porsche introduced the 911. It was first shown in 1963 as the 901, but when production got under way in 1964 the name had been changed, because of complaints from Peugeot, who had trade mark rights to all such numbers with '0' in the middle!

Like the 356, the 911 had a pressed steel monocoque body/chassis design, and a rear-mounted air-cooled engine, but the wheelbase was four inches longer than before, there was genuine 2+2 seating, and the engine was a 1991cc flat-six design producing 130hp at first. Styling was by Butzi Porsche, the third generation of this remarkable family, while engine development had been led by Ferdinand Piech, who was another grandson of the founder.

Compared with the old car, there were many chassis changes, for the front suspension was now by an intriguing mixture of MacPherson strut linkage, but longitudinal torsion bars as the springs, while rear suspension was by semi-trailing arms, and transverse torsion bars. There was rack and pinion steering, disk brakes all round, and a five-speed, all-synchromesh gearbox, mounted ahead of the engine, and the line of the rear wheels.

Although early versions suffered from inadequate roadholding, partly due to the

fitment of narrow wheels and tires, the type was gradually but successfully improved. Indeed, the whole, long-running, story of the 911 family is one of the diligent development of a basic theme. The first major up-grading came in the summer of 1966, when the 160hp 911S was launched. This car had ventilated disk brakes, five-spoke alloy road wheels, and a top speed of at least 140mph.

At the same time, the first derivative from the original fastback coupe body style was announced, this being an openable-top car known as the Targa. Unlike normal convertibles, it had a permanent roll-protection bar fixed in a loop, above and behind the front seats. The term 'Targa' has now been adopted all over the motoring world, to describe this sort of car.

In 1967, Porsche astonished the world by offering a semi-automatic type of transmission known as 'Sportmatic,' in which there was a torque converter and a clutch, but clutch control was by a microswitch in the gear lever, which still had to be used to change ratios.

In 1968, for what has become known as the B-Series cars, the car's wheelbase was extended by two inches by moving the

Left: 1972 2.3-liter flat six Porsche 916 Prototype. Only 11 were built.
Below left: Porsche 911, 1979 Monte Carlo rally.

rear wheels back, relative to the floorpan and engine/transmission position. The engine's crankcase was to be made from magnesium, rather than aluminum, and the 911S was given six inch wide wheel rims. A year later the S and 911E engine was given Bosch fuel injection instead of multiple carburetors, and at the same time a 110hp carbureted 911T joined the range as a lower-price version to close the gap left by the demise of the four-cylinder 912.

The C-Series cars, announced in summer 1969, had an engine capacity enlarged to 2195cc, and two years later the E-Series cars appeared, with even larger, 2341cc units. In each case, this was done to increase low speed torque, to make the cars more 'driveable', and to optimize them to meet USA exhaust emission rules.

In summer 1972, F-Series heralded the revival of the Carrera name, for cars with 2687cc, 210hp at 6300rmp, even wider (seven inch) rear wheel rims, flared wheel arches, a duck-tail spoiler, and a great competition potential. This car could accelerate to 60mph in a mere 5.5 seconds, and could achieve 150mph.

This enlarged engine was fitted to all cars from the summer of 1973, except that the Carrera stayed one jump ahead by being given a 2993cc/230hp unit. All cars now had Bosch fuel injection, and there were important detail style changes involving much heftier front and rear bumpers.

From October 1974, the motoring world could lust after the mighty 911 Turbo (which, in factory parlance, was more properly the 930 Turbo). This car had a body style similar to the race-modified Carrera RSR, with even more outrageous wheel-arch flares, and a larger 'tea-tray' rear spoiler. The 2993cc unit had a KKK turbocharger allied to Bosch K-Jetronic fuel injection, and the 260hp power output was transmitted via 215-section Pirelli P7 tires.

Quite simply, the Turbo set new supercar standards, and from August 1977 it was made even more outstanding by having an even larger 3299cc engine, and 300hp at 550rpm. The Pirelli tires were yet larger – of 225 section – and the top speed was over 160mph.

In summer 1975, the 911S and 911SC 2.7-liter models gave way to the re-developed cars, which had the 3-liter power unit and 200hp, and were titled Carrera 3. Two years later there was a further naming somersault, with the Carrera name dropped, and the 911SC title revived, while a full convertible model was finally made available in 1982 (and called Cabriolet), all those years after the Targa had been put on sale. The latest development, for the start of the 1984 model year, was that the 'base' engine was once again enlarged, to 3164cc, and its peak

Above: 1971 Porsche 911E.
Below: Porsche 911 Cabriolet still shows styling first developed in the early fifties and still commands high prices.

THE 928S

power raised to 231hp at 5900rpm, at which point the cars were once again renamed Carrera!

We must now revert to the mid-1960s. When the old Type 356C cars were finally dropped, there was still a demand for a lower-priced, less-sophisticated, car than the new 911, and in 1965 this was met by the Type 912. Effectively, this was a 911 body/chassis and suspensions, but fitted with the flat-four 1582cc engine of the obsolete 356. By 911 standards, performance was modest, but a total of 30,300 were sold until the car was superseded, partly by the 911T, and partly by the Type 914.

The 914 was yet another mould-breaking design, intended to be a lower cost range than the 911, using many mass-production parts. Because the mid-engined layout had proved to be so successful in motor sport, this was chosen for the new car, which is to say that the engine was ahead of the line of the rear wheels, with the transmission behind it. At the same time it was discovered that VW wanted a boost to their image, so it became a joint venture, the name became VW-Porsche, and a VW engine and transmission was chosen.

Left: Water-cooled V8 engine was something of a departure for the Porsche 928S.
Below: Porsche 928.

The structure was a conventional steel monocoque, with distinctive (some thought strange) styling. Front suspension was from the 911, but the semi-trailing rear suspension was linked to coil springs. At first, the car had a 1679cc air-cooled flat four (from the VW411), but from 1972 this was enlarged to the full 1971cc size, with fuel injection. A total of 115, 646 were made up to 1975, when it gave way to the 924.

A Porsche version of this car, which had no VW part in its title, used the 911T's 1991cc engine, when the car was known as 914/6, but this was not a success. Only 3,360 were built, and it had to be dropped in the summer of 1972.

As another stop-gap, therefore, the 912 model was re-introduced for sale only in the USA, for 1976, this time with the 2-liter flat-four, but it was a short-lived marketing effort.

The really important car of 1975 was not the re-introduced 912, but the 924, which was a complete design break with all established Porsche traditions. Yet again, here was a car intended as a cheap Porsche, and like its predecessors it used many VW Group (or, to be more precise, Audi) parts. However, not only was it a front engined car (with a gearbox in unit with the final drive), but it had a water-cooled engine!

The new layout was much more logical for the safety conscious mid-1970s, for it gave more predictable roadholding, and better occupant protection in crashes. The engine was linked to the transmission by a large diameter alloy torque tube. The engine itself was a 2-liter, single overhead camshaft conversion, of the previous 1.9-liter Audi 100 four-cylinder unit, and would later be used in the VW LT van, and in the American AMC Gremlin car. With Bosch K-Jetronic fuel injection, there was 125hp for Europe, but a mere 95hp for USA sales.

Though early models were criticized for being too noisy, and too basic, the styling was always considered pleasant, if conservative. Porsche struck a production deal with VW/Audi, whereby assembly would take place at the ex-NSU factory some little distance from Stuttgart.

The 924 Turbo was offered in 1979, and was a considerable improvement over the basic model, for it had a 177hp engine. Externally, it was distinguishable by the extra cooling vents in the nose, and the wrap-around spoiler on the tail. A limited-production model of the same period was the 924 Carrera GT, which was a modified Turbo, with flared wheel arches, and even more power (210hp), for racing, and occasional road use.

From 1981, however, there was the 944 model, which was basically a further-developed 924, fitted with a new Porsche-designed slant-four engine, with a single overhead camshaft, and 163hp from its 2479cc. This, the enthusiasts said, was what the real 924 should always have been, and with its enhanced performance and smoothly shaped wider wheel arches it soon became very successful.

In the meantime, the 928 had arrived in 1977, as Porsches latest idea of the modern supercar. Like the 924, it had a front-mounted, water-cooled, engine, and rear gearbox, and broadly similar styling outlines, but it was all-new, and totally Porsche.

Left: 1984 Porsche 924 Lux Coupe. *Below left and below:* 1984 Porsche 944 Lux; the 924, developed for VW, gets proper Porsche power and looks.

Designed to attack the lucrative expensive coupe market sector, it was smooth, silent, extremely efficient, but somehow without much character as first revealed. The rounded styling, with a large expanse of glass, and wide hips, was not at all dramatic.

Its engine was a new light-alloy V8 unit of 4474cc, with single overhead camshafts per bank, and the inevitable Bosch K-Jetronic fuel injection, the power output being 240hp at 5250rpm. There was a choice of transmission – a five-speed manual unit, or a three-speed automatic, actually bought in from Mercedes-Benz.

The original 928 was actually slower than the 911, and because it was heavier it also had a worse fuel consumption. Sales were slow at first, so from 1979 it was joined by the 928S, in which the engine size was enlarged to 4664cc, the power output ot 300hp, and new aerodynamic aids were added, notably a wrap-around rear spoiler under the tailgate glass. This car could breach the important 150mph market, and rocket up to 60mph in around 6.5 seconds. The 928S2 which followed, had 310hp, and even more performance. These cars, of course, were not at all like the 911s, which were still being made, for they had vice-free handling, close-coupled four-seater accommodation, and full air-conditioning as standard equipment.

Riley

Above: 1922 Riley 1496cc Sandracer.

Riley was one of Britain's first car manufacturers, and like others they had begun by making bicycles, while even earlier they were involved in the weaving trade. The very first Riley car was built in 1898, and set the trend for the future by being full of innovations.

Among the technical 'firsts' later credited to the company are the mechanically-operated inlet valve (1898), adjustable front seats and steering column (1919), and detachable wire-spoke wheels (1907). Indeed, demand for the Riley wheel was such that Rolls-Royce took out a licence to manufacture it, along with at least eight other manufacturers. It is said that the Riley family almost gave up making cars to concentrate on the wheels business.

Riley always had an extremely inventive design department, seeming to announce more cars than they actually made. The first truly well-known Riley was the Eleven of 1919, a vehicle fitted with all manner of bodies, including two-seater and four-seater sports models. In 1924 this type reached its development peak as the short-chassis Redwing, the name being because of the characteristic scheme of a polished alloy body shell, with red wings. It was fitted with a 1498cc engine.

In 1926, Riley exhibited the new Nine at the London Motor Show, and this was the model which was to give the company a real sporting impetus. The heart of the car was a completely new engine, designed by Percy Riley. Technically advanced, it had twin high camshafts, one on each side of the block, operating short pushrods to the opposed overhead valves, which were in part-spherical combustion chambers. This was an efficient breathing layout, which Riley were to retain for the next 30 years.

As introduced, this was a 1087cc engine, which produced 32hp at 5000rpm, but an early disadvantage was that the crankshaft only had two main bearings. This failing was soon rectified, and later engines had up to 50hp.

The early Nines were fitted with saloon or tourer bodywork, then in 1927 a sports version, the Brooklands, named after the famous Surrey racetrack with banked circuit, was announced. This derivative had been produced with the help of Parry Thomas and Reid Railton, both famous in motor sport at the time, and actually built by Thomson and Taylor inside the track itself. Priced at £395, it had an extremely low build, the seated driver being able to touch the ground, outside the bodywork, with his

Below: 1935 Riley 1½-liter Sprite prototype.

Above: Six-cylinder 1498cc Riley Sprite from 1936.

hand. Racing successes started to arrive with this car, or its developments, including third place in the Brooklands Double Twelve hour race of 1930, and second place in the Irish Grand Prix the same year.

By 1930, too, there was a super sports version of the Nine, called the Ulster Brooklands, which had a balanced engine, and extra carburetors. Further competition successes followed, including outright victory in the 1932 Tourist Trophy, and a class win at Le Mans in 1935.

A variety of names were given to the sporting Nines, including the £298 Gamecock of 1932, and the Lincock fixed head coupe. The Imp, built on a short-wheelbase chassis, was the last of the sporting Nines, and built in 1934. This had a most attractive body style, with flared mudguards, offered 75mph performance for just £325 – 75 examples were produced.

For racing, incidentally, the highest output ever recorded for a three-bearing Nine engine was 183hp at 7400rpm, naturally with a lot of supercharger boost.

Riley introduced an in-line six-cylinder, with much the same type of valve gear and cylinder head breathing as the Nine, in 1928, and although this was only fitted to a limited number of sportscars, it also found a home, much changed, in the famous ERA racing cars of 1934 onwards. Virtually a six-cylinder version of the four-cylinder Nine, it had three main bearings, and was originally built as a 1633cc unit. The most striking of all was the MPH model, which looked very similar indeed to the Imp, but this was expensive, and only 12 such cars were made.

Victor Riley never believed that this engine could be highly tuned, at first, but was eventually proved wrong, not only in factory competition cars, but in Raymond Mays' famous White Riley, and the ERAs which evolved directly from it. These engines, drastically modified, and with supercharging, were built in 1.1, 1.5 and 2.0-liter form, the most powerful of which certainly had more than 300hp at the peak of their development. In semi-private hands, the six-cylinder engine, usually with a carburetor to each cylinder, was amazingly successful in the hands of Freddie Dixon.

A new four-cylinder engine of 1496cc –

Below: 1935 Riley Imp 9hp derived from the earlier Brooklands model.

129
ELM 516

the 12/4 – was announced in 1934, this following classic 'Nine' lines, but being the work of Hugh Rose, who later moved to Lea Francis, also of Coventry, and designed a very similar engine for them. The classic layout had camshaft drive by chain at first, but gear drive in due course, and was used in the Kestrel saloon, and then in the Sprite. The Sprite of 1935 looked just like the MPH, but had the four-cylinder engine, and a choice of traditional Riley, or 'fencer's mask' radiator grille. By this time, however, the BMW 328 had arrived on the motoring scene, which explains why a six-cylinder engine was also made available. Cars known as TT Sprites were very successful in sportscar racing.

By 1937, however, Riley faced financial ruin, said to be because of the diversity of their model range, and early in 1938 they called in the Receiver. Then, in the autumn of that year, Lord Nuffield took over the company, and weeks later sold it at a considerable personal loss to his Nuffield organization, where it joined Morris, Wolseley and MG.

Although Riley, as a marque, lingered on for Nuffield, and BMC, until the end of the 1960s, no more sporting models were ever put on sale. However, the engine was also used in most Healey cars built between 1946 and 1954.

Top left: 1936 Riley Sprite 1498cc.
Top right: Riley RM convertible.
Bottom: 1930 Riley 9hp Brooklands low-chassis two-seater. The nine had taken part when the classic RAC TT road race was revived in 1928.

Stutz

The archetypal early American sportscar – or roadster, as it was then known – was the Stutz Bearcat, which competed directly with the Mercer Raceabout of the day. Harry C Stutz had worked for J N Willys at Marion, and for other motor car manufacturers, before leaving to produce combined gearbox/final drive transaxles, which were bought by several concerns.

In 1911, however, this company built a successful Indianapolis racing car to publicize its wares, and followed up by launching a passenger car. Their greatest fame came from the Bearcat model, announced in 1914, built before the company completely changed its direction in the mid-1920s.

The Bearcat was a large two-seater, with a long chunky bonnet, and no weather protection for the occupants. There was a 60hp, four-cylinder, Wisconsin engine up front, which produced massive torque, and gave the car a lively performance. The style of the car epitomized speed, for it had two bucket seats, a cylindrical fuel tank behind those seats, a well angled steering column, and all the controls and levers completely exposed.

Houk wire wheels were made standard from 1917, though the traditional wooden wheels could still be ordered. In view of Harry Stutz's expertise with rear-mounted gearboxes, it was not surprising that the Bearcats had this feature, which helped to balance the weight distribution, and improved the roadholding.

Stutz himself profited mightily from Bearcat sales, then sold out his stake in the company in 1917. With this money he started up another car company, HCS, but this did not prosper. Although he remained connected with motor vehicles for many years afterwards, these were not cars, but mostly Stutz fire engines. He died in 1930.

The Stutz Motor Car Co. of America, building cars at Indianapolis, revised their sportscars, with conventional gearboxes close to the engines, during 1921. Stutz's own design of engine – in four-cylinder or six-cylinder guises, had already been made available, from 1918. The 'four' was a 360cu. in./5.9-liter side valve unit, which developed 88hp, while the 'six' had overhead valves, and a similar power output, but only displaced 268cu. in./4.4-liters.

A typical Bearcat of 1921, incidentally, had electric starting, rear wheel brakes, a multi-plate clutch and a three speed transmission, shock absorbers, a spotlight mounted in the windscreen pillar, and rather more enclosed bodywork than the original examples, even with a vestigial hood. The final versions of the car, in the mid-1920s, which had the sporting six-cylinder engine, were known as the Speedway Six, while the Bearcat itself was called the Speedway Four.

The firm's peak year had been in 1919,

Right and below: 1914 Stutz Roadster, with 6.5-liter, four-cylinder engine producing 60hp at 1500rpm, was the first of Harry C Stutz's cars to be called 'Bearcat'.

when 8500 cars were built, and a year later C W Schwab (president of Bethlehem Steel) bought the business. Ever after that it was in the doldrums, losing $500,000 in 1924 alone. To rectify matters Schwab hired F E Morkovics (from Marmon) as President, Charles Greuter was brought in to produce a new European-style sportscar, and $1 million was made available for tooling.

The new car appeared in 1926, as the AA, or Vertical Eight, with a straight-eight cylinder 289cu. in./4.7-liter engine. Marketed as the Safety Stutz, it had advanced four wheel hydraulic brakes, and used wire-reinforced glass in the windscreen. A speedster version of 1927 was given the name Black Hawk. Stutz cars of this period performed well in American racing, and one car even finished second at Le Mans in 1929, behind the winning 4½-liter Bentley.

The Bearcat name was retained for short wheelbase versions of subsequent luxury models, but all car production ended in 1935. Many years later, in 1970, a new Stutz company offered GM-based replicars from New York, and these still use the Bearcat title.

Sunbeam

Until the mid-1920s, Sunbeam was one of the very few British companies to seriously build, develop and race a team of Grand Prix cars. By winning the French GP in 1923, Sir Henry Segrave became the first British driver to win a Grand Prix in a British car, a feat which was not to be repeated until the 1950s. However, although Sunbeams had a fine reputation, there was no sportscar for road use while the company was upholding British prestige on the race track.

It was not until 1925 that the 3-liter model appeared. Sunbeam, which was a part of the Sunbeam-Talbot-Darracq combine, wanted to produce a car superior to the 3-liter Bentley (which had won the Le Mans 24 Hour race in 1924). The 3-liter was fitted with an engine based loosely on that which won the Grand Prix events, for it had twin overhead camshafts, driven by gears, while the camshaft and crankshaft were superimposed in seven main bearings. Lubrication was by dry sump, and carburation was by twin Claudel-Hobson instruments. A four-speed close-ratio gearbox was fitted.

A peculiarity of the four-seater open 3-liter was that it had close-fitting cycle type front wings, which turned with the wheels. This car was never built in quantity. Only 250 were produced before manufacture ceased in 1930 – one reason being that the chassis got a reputation for being too frail, the other being that it was rather too complicated a design. Perhaps if there had been a more rigid, short-chassis, type, Sunbeam might have sold more of these cars.

The company finally went into liquidation in 1935, as the S-T-D combine collapsed from the top, and the Wolverhampton-based company was eventually taken over by the Rootes Group, along with its sister company of Talbot, to join the Hillman and Humber marques.

Sportscars as such did not figure again behind this badge until 1953, when the two-seater Alpine was announced. This was a Rootes car, pure and simple, being based on the chassis and running gear of the Sunbeam-Talbot 90 saloon. As a sportscar its frame was stiffened up with extra side plates, and the nose of the car was exactly like that of the saloon. There was independent front suspension by coil springs and wishbones, while the rigid axle was located by half-elliptic leaf springs. The four-cylinder 2267cc engine developed 80hp at 4200rpm, and there was a four-speed gearbox operated, as was the fashion of the day, by a steering-column change. Theoretically this was to allow three people to sit in the front seat of all such cars, but seat contours in the Alpine made this impossible. On announcement the car was priced at £1269, compared – say – with £780 for the MG TF of the day.

The car was quite successful in competitions, and to celebrate its announcement, Sheila van Damm drove one at 120mph on the Jabbeke Autoroute in Belgium. In the same year the team cars won four Coupes des Alpes in the Alpine rally. Using Alpines in 1953 and 1954, Stirling Moss became only the second driver in history to win a Coupe d'Or for three consecutive 'clean' runs on this difficult event. The car was discontinued in 1955.

In the same year Rootes announced a new sports saloon, which they called the Sunbeam Rapier, and from mid-1959 a sportscar derivative of this car appeared, once again called the Alpine. Its roots were complex, for the Series III Rapier's running gear and suspension were fitted to a shortened version of the underpan, and topped off by a smart two-seater body shell with prominent tail fins, and wind-up windows.

At first the Alpine's engine size was 1494cc, and the car could reach almost 100mph, but it took about 13 seconds to accelerate from rest to 60mph. Over the years the car always suffered a little from having a smaller engine size than its main competitors – the MGA/MGB and the Triumph TR Series, but it made up for this in styling, and equipment. The price was always right – on announcement it cost £971, compared with £940 for the MGA 1600.

Right: 1926 Sunbeam 3-liter. After years of top-class racing, Sunbeam retired from the track in 1925.
Far right: 1928 Sunbeam 'long 25hp'.
Below: 1963 Sunbeam Alpine.

Below: Sunbeam 4.3-liter V8 convertible made a real performer from a saloon-derived 2-seater.

LUCAS
ESSO EXTRA
85
MXP 535

XV 749

55
9202 RW

Alpines were built from 1959 to 1968, growing up to Series V, and to a 1725cc engine. Overdrive transmission was always optional, automatic transmission was available for a short time, and the cars could be supplied as open sports or closed hardtop cars. A few fastback conversions were produced, with Rootes approval, by Harrington, of Sussex. This type of Alpine had a limited, but creditable, competitions career, which included winning the Index of Thermal Efficiency award at Le Mans on the first of three visits, in 1961.

The Alpine's basic lack of performance was rectified, in 1964, in no uncertain manner. At this time, competitions manager Norman Garrad's son, Ian, was Rootes's Californian representative, and was well aware of the way that Carroll Shelby had transformed the AC Ace into the Cobra with a Ford V8 engine transplant.

With the help of Ken Miles, a similar transplant was carried out, for a 4.2-liter/164hp Ford V8 engine was fitted to an Alpine, the prototype was shipped to the UK, productionized, and shortly announced as the new Tiger. Assembly of Tigers, incidentally, was always by Jensen at West Bromwich.

There were no visual styling changes, for the engine slipped under the bonnet without any skin panels needing to be altered. The body structure, too, was very strong, and accepted the big increase in torque without problems. The transformation in performance was remarkable. Whereas the current Alpine needed 14 seconds to reach 60mph from rest, the Tiger took only 9.5 seconds, while the maximum speed rose from 95/100mph to 118mph. The UK price of the car was £1445.

The factory made one attempt to turn this car into a race-winner, when a team started the 1964 Le Mans race (both cars retired with engine failures), but were more successful in rallies. Peter Harper's car was sixth at Monte Carlo in 1965, and set best performance in the Alpine, before being disqualified after the finish on a technical detail.

Production of Alpines ended in 1967, soon after the 4.7-liter Series 2 model had been put on sale, for by this time Rootes had been absorbed by Chrysler, who could not stomach producing a sportscar with a rival manufacturer's engine.

Right and far right: 2.3-liter Sunbeam Alpine convertible, 1955-60.
Below: 1961 Sunbeam Alpine Series II.
Below right: Sunbeam Tiger 4.3-liter V8 convertible.

520 NOU

SUNBEAM
899 KYD

Talbot

The famous series of 'Roesch' Talbots (so named after then Swiss-born designer) did not arrive until the 1920s, by which time the marque had already enjoyed twenty successful years. The company was originally founded by the Earl of Shrewsbury & Talbot (hence the name), to assemble cars in London, from French components, but from 1905 totally British types replaced the original Clement-Bayard types.

Amongst these first true Talbots was a four-cylinder 25hp model, introduced in 1908. An example of this car, carefully prepared, and stripped out for testing, was the first-ever car to cover 100 miles in an hour, on the Brooklands track. Percy Lambert was the driver.

During World War I, Talbot made staff cars and ambulances, and in 1916 they also gained the services of Georges Roesch as their chief engineer. Aged 22, Roesch had already worked for Gregoire, Delaunay-Belleville, Daimler and Renault.

After the war, Talbot merged with Darracq and Sunbeam, becoming a part of the S-T-D combine, and this led to Roesch becoming subservient to Sunbeam's noted chief engineer, Louis Coatalen. The result was that Talbot were forced to produce some dull models, though Roesch's first minor triumph was to create the Talbot 10/23 from a little Darracq model.

Finally, in the mid-1920s, Roesch was given the freedom to design a brand new Talbot. His remarkable, dynasty-founding, 14/45 six-cylinder model first appeared at the 1926 London motor show. Although its engine displaced only 1666cc, it was a highly efficient unit, with overhead valve gear, whose rockergear was located on instantly adjustable knife-edges. There was a very light flywheel (which also incorporated a cooling fan), and this helped speedy throttle response.

Even though it was a substantial car, the 14/45 was an overnight success, for it could exceed 60mph when fitted with saloon car bodywork. Not only that, but it laid the foundations for a succession of larger-engined Talbots.

The first of these was the 75 (with a 2276cc engine), but the first of the truly sporting Talbots was the 90, whose engine developed 93hp at 5000rpm, in road trim, with the same engine capacity as the 75. As prepared for racing by Fox and Nicholl Ltd., these cars enjoyed numerous racing successes, the most notable of which was to finish third and fourth, at Le Mans, behind the two 6½-liter 'Speed Six' Bentleys. This, indeed, was a recurring Talbot problem, for although their cars habitually performed well, they were usually beaten to ultimate success by cars with much larger engines.

At the end of 1930, Roesch developed an engine which was still based on the same overall dimensions and layout as the 14/45 unit, yet had a capacity of 2960cc. In touring trim, this unit produced 100hp, but for racing, using petrol-benzole fuel mixture, it had up to 140hp. This new model, named the 105, took third places at Le Mans in 1931, and again in 1932. In 1932, also, a team of three cars lost no penalties at all in the tough Alpine trial, so winning the Coupe des Alpes for the marque.

Right: 1933 Roesch Talbot 90 Vanden Plas, one of Georges Roesch's last designs for the company.
Below left: Talbot 90.
Below right: Talbot 105.

GX 68

38

BGM 23

In 1934, the ultimate derivative of the Roesch theme arrived, this being the 3.37-liter 110, which was still based on the same engine layout. This mighty unit developed 123hp in standard form, or up to 170hp for racing. A compact, and immaculately prepared, four-seater open version, once lapped Brooklands at almost 130mph, making it the very fastest 'touring car' on the famous banked track.

Unfortunately Talbot, which was a profitable company, could not survive alone, so when Darracq and Sunbeam collapsed, having been hit by the Depression (and also by Coatalen's over-spending on motor racing), Talbot were dragged down with them. Almost 12,000 of the 14/45 and modernized 65 models had been sold, together with appreciable numbers of the larger-engined cars.

Talbot was speedily bought up by the Rootes Group, who also purchased Sunbeam later in the same year of 1935. Initially the new owners carried on building Roesch-designed cars, but as the parts ran out, Humber components and engines were used instead. Roesch left in disgust.

In 1938 the famous marque fizzled out, and to the dismay of purists, Sunbeam and Talbot names were then combined by Rootes, to be used on upmarket versions of Hillman and Humber cars. After World War II, quite sporting Sunbeam-Talbot 90s appeared, but from 1954 the Talbot name was dropped once again. Successors to those cars, such as the Alpines and Tigers, were Sunbeams only.

Further to confuse the story, the old Darracq concern continued to make its own breed of sporting 'Talbots' in France, after Tony Lago acquired the business and both users of this name ultimately ended up in the same, Peugeot-Citreon, combine from 1978. Hence the revival of the Talbot name on the French and British made cars from 1979.

Right: 1937 Talbot 105 drophead tourer, created after the S-T-D combine had been broken up.
Below: 1934 Talbot 105.

De Tomaso

After an undistinguished spell driving a variety of racing cars in his native Argentina, Alejandro de Tomaso moved to Italy. Here, for a time, he worked for the Maserati brothers at their OSCA concern, before leaving to set up on his own, and build racing cars.

Having watched, and been impressed by, the sports racing Cooper of the late 1950s, he became convinced of the virtues of the mid-engined configuration. After building a series of junior Formula single-seater racing cars, he produced his first road car, the Vallelunga, which had a mid-mounted Ford Cortina GT type of four-cylinder engine, a backbone chassis frame, and all-independent suspension.

Later in the 1960s, when the AC Cobras were at the height of their fame, de Tomaso developed and announced the Mangusta. This was the Italian name for a mongoose, the only other animal fast enough to catch and eat cobras! Like the previous Vallelunga, the Mangusta had a backbone frame, with double wishbone independent front and rear suspensions. For this car he chose to use an American V8 engine, in fact a 5.0-liter Ford unit which produced 305hp, and drove through a five-speed ZF transaxle. The Mangusta had sparkling performance, for it could spring to 60mph in 6.5 seconds, and to 100mph in 14.5.

The body shell, mainly constructed of steel, had been styled by Giugiaro in his time at Ghia, and was unusual in that access to the engine bay (behind the seats) was gained bylifting roof panels, gullwing style, from a central spine. Over 400 Mangustas were produced between 1967 and 1972.

In 1967 de Tomaso acquired Ghia, and in due course he sold that concern to Ford of Detroit, who also took a stake in his own business. At the same time, they agreed to take over the marketing of de Tomaso products in North America.

The result of this liaison was the new

Right and below: 1971 De Tomaso Mangusta, styled by Giugiaro and first seen at Turin in 1966, used 302 V8 Ford Windsor engine and could reach 150mph.

Pantera (which was Italian for panther), first seen at the New York Show of 1970. In many ways the layout was similar to that of the Mangusta, except that it had a combined steel body/chassis unit. It was a truly international car in every way, with a US Ford V8 engine, German transmission, British brakes and steering, French tires, Italian body and assembly.

The engine was Ford's latest 351cu. in./ 5.7-liter 'Cleveland' unit which, when fitted to the HO (High Output) Pantera gave it a 0-60mph time of 7.0 seconds, and 0-100mph in 15 seconds. The European version of the car, unencumbered by exhaust emission equipment, could exceed 150mph. As with the Mangusta, there was coil spring independent suspension all round, Girling ventilated disks at front and rear, and the cast magnesium wheels were shod with Michelin XVR radial ply tires.

In 1973 a 'GTS' version was introduced for European markets, and with a 350hp 'Cleveland' engine top speed was claimed to be 175mph. By 1974, however, a combination of factors had killed Pantera sales in the USA. The cars rusted badly, and were unreliable, while the handling was somewhat suspect. The very high price, $11,061, allied to the drop in large-engined car sales after the 1973 energy crisis, didn't help.

Ford therefore gave up its de Tomaso links, but retained Ghia. The Pantera, along with other de Tomaso models, carried on being built, in very small numbers, until the mid 1980s.

Right and far right: De Tomaso Pantera, styled by Tom Tjaarda, used the 351 Ford Cleveland V8 engine.
Below: The later De Tomaso Pantera GTS.

DETOMASO
PANTERA
GTS

E E 94285
GTS

Triumph

The Triumph car concern grew out of the successful motorcycle business, but the motor cycles were more famous for many years. The original Triumph car was announced in 1923, but it was not until the 1930s that the first sporting model, the Southern Cross, was announced.

This car had a conventional chassis, and was fitted with a Coventry-Climax four-cylinder engine of 1018cc, having overhead inlet and side exhaust valves. This was later enlarged to 1122cc. From 1934, too, there was a new generation Southern Cross, this time on the more advanced Gloria chassis, and this was fitted with four-cylinder 1232cc or six-cylinder 1991cc Coventry-Climax style engines. Derivatives of these cars were successful in rallies, particularly in Donald Healey's hands.

The most notorious pre-war Triumph sportscar was the eight-cylinder Dolomite, of which only three examples were built in 1934/35. This car was an acknowledged copy of the Alfa Romeo 8C2300 model, and had an almost identical-looking engine, with a supercharger, though it displaced 1990cc, and produced 140hp. It was to have been very expensive – £1225 – but the company could not afford to put it into production, and it was killed off within months.

After going bankrupt in 1939, Triumph was finally bought up by Standard in 1944, and in 1946 the first 'Standard-Triumph', the 1800 Roadster, appeared. This car, aimed at the export market, had three-abreast seating, a steering column gear change, and was the last car in the world also to offer a fold-up 'dickey' seat behind the driver. At

first it was fitted with a 65hp 1776cc engine, exactly as supplied by Standard for the Jaguar 1½-liter models, but from 1948 it was given the 2088cc/68hp Standard Vanguard engine, and was called the 2000 Roadster.

In the early 1950s, Standard saw a huge gap in the vast US market for a sportscar between the MG TD of 1250cc, and the Jaguar XK120 of 3442cc, and with this in mind, a new sportscar was designed around a modified Vanguard engine and gearbox. The aim was to produce a 90mph car which could be sold for £500 (basic).

A prototype, retrospectively called the TR1, was at the London Motor Show of 1952, but soon afterwards this was found to have poor roadholding and performance. A crash re-design programme was carried out, and the definitive TR2 was ready to go on sale in summer 1953. The original car had used a modified Standard Flying Nine

Left: 1959 Triumph TR3A.
Below left: Triumph Gloria Southern Cross. Precious few of these cars still survive.
Below: Triumph TR2 and TR7 (foreground).

frame, and a 75hp engine, with an exposed spare wheel on the tail. The TR2, however, had its own special frame (with coil spring independent front suspension), an engine further tuned to 90hp, and a restyled, longer, tail. Its reputation was founded on competition success. Right from the start a development car achieved 124mph on the Jabbeke autoroute in May 1953.

The TR3 was introduced in October 1955, with modifications which included a front grille in the air intake, and a 95hp engine. As already developed for the TR2, overdrive, wire wheels and a hardtop were optional extras. A year later the TR3 broke new ground by becoming the first series production British car to have disk brakes – Girlings on the front wheels. Then, from January 1958, came the final style change, when the TR3A was launched, complete with a full-width front grille, and a 100hp engine (as for the last of the TR3s). A 2138cc engine also become optional for 1959.

At Earls Court in 1961, there was a completely new body shell, styled by Michelotti, for the TR4, this having wind-up door windows, and better appointments. The chassis and running gear was modified TR3A, except that wheel tracks were wider, and the 2138cc engine was standard. Top speed was about 109mph, and 0-60mph acceleration took about 11 seconds. The old model, modified and renamed TR3B, remained in production for another year.

Next, early in 1965, came a new chassis frame, incorporating coil spring independent rear suspension, still with the 2138cc engine, this being the TR4A. The rear end was the same basic layout as used in the Triumph 2000 saloon, and the new car had a much softer ride than before. For the USA only, it was possible to order a TR4A with a rigid rear axle as an alternative.

Externally, the TR4A was nearly identical to the TR4, except for styling details, which included the use of small indicator lights at the front of chrome strips along the flanks, and a new grille. The TR4A had 140hp, and had a top speed of almost 110mph, just marginally faster than before.

The big change came in the summer of 1967, when the old four-cylinder engine was dropped. The TR5, still using the same basic chassis, suspension, and body shell as before, was given a fuel-injected 2498cc six-cylinder engine, which had evolved from the Triumph 2000 saloon's unit. In this form it had 150hp, but for the USA, with carburetors, and anti-emission controls, only 105hp was available. With the most powerful engine (which was detuned, in later years, incidentally), up to 120mph could be reached.

As the TRs gradually became larger, faster and more sophisticated, Triumph made space for a new smaller sportscar, the Spitfire, which was also a competitor for BMC's Sprite/Midget models. Although it had a unique steel backbone-type chassis frame, and a Michelotti-styled two-seater body style, most components were modified Herald items, including the engine and

transmission, the all-independent suspension, and the steering gear.

The first car had 63hp from 1147cc, and a top speed of 90mph, and was soon available with overdrive, wire wheels, and a detachable hardtop, all as options. In the spring of 1965, the Mk 2 Spitfire was launched, with 67hp, and from the spring of 1967 there was the Mk 3, with 1296cc, 75hp, and near-100mph performance.

But that was not all. At the end of 1970, the Spitfire became Mk IV, with a different, swing-spring, type of rear suspension which gave greatly improved road holding, and with a restyled body shell which included a cut-off tail, and a more angular optional hardtop. The final derivative, produced from late 1974 to 1980, had a 1493cc engine, and a genuine 100mph top speed in European tune. All in all the Spitfire was extremely successful, outselling the 'Spridget' nearly every year – a total of 314,342 were built in 18 years.

The GT6, first seen in 1966, was essentially a Spitfire structure, modified with a smart fastback/hatchback top, and with a 1998cc six-cylinder engine of Triumph 2000 type, closely related to that of the TR5/TR250. Many people compared it to a scaled down Jaguar E-Type. With 95hp, it was a 100mph-plus car, but the roadholding was suspect. The Mk 2 of 1968 was a much improved machine, with 104hp, and a revised type of rear suspension having a wishbone linkage. The Mk 3, introduced in the autumn of 1970, had a restyled skin, like that of the Spitfire Mk IV, with cut off tail. Finally, for 1973, the GT6 reverted to the swing spring rear suspension, but it was dropped at the end of the year. A total of 40.926 GT6s were produced.

After just 2947 TR5s and 8484 TR250s (USA only) had been produced, these models were superseded by the TR6. Mechanically there were no changes, and the basic body shell was as before, but the front and rear had cleverly been modernized by Karmann of West Germany. At first the UK-market TR6 cost £1339, while the USA-market version, also called TR6, cost $3375, slightly cheaper than before. The TR6 model ran on, and changed only to keep abreast of USA legislation. With a few cosmetic improvements it continued until the summer of 1976: a total of 94,619.

The TR7 of January 1975 was always a controversial car, and was to be the final member of the long-running TR family. It was completely different from its forebears, in having a unit construction body/chassis structure and – for the first four years – was only sold in notchback two-seater coupe style. The wedgenose style, with concealed headlamps, was very different from previous TRs, as was the use of a 1998cc single-overhead-cam four-cylinder engine (a modified Dolomite saloon unit). Independent front suspension was by MacPherson strut, but there was a rigid rear axle, on coil springs, with radius arm location.

Compared with the TR6, it was much

Above left: 1972 Triumph TR6.
Above: 1970 Triumph Spitfire.
Below: 1980 Triumph TR8.

smaller, but had greatly improved areas of comfort, handling, roadholding and economy. At first it had a four-speed all-synchromesh gearbox, but from 1976 there was the option of automatic transmission (the first time ever on a TR), or a more robust five-speed gearbox, actually that fitted to the new big Rover hatchbacks, and the latest Series III Jaguars. Top speed, with 105hp, was 109mph, 0-60mph took 9.1 seconds, and fuel consumption sometimes approached 29mpg.

A convertible TR7 was introduced in 1979, and the style was more pleasing than the coupe, which also continued. From 1978, BMC started rallying a Rover V8 (3528cc) engined version of the car, called the TR7 V8, but the true TR8 production car was only sold in the USA in 1980 and 1981. Other versions, including a 16-valve 'Sprint' derivative, were killed off by a long strike, which paralyzed the Liverpool production factory for months in the winter of 1977/78. Production was later transferred to Coventry, and finally to the Rover plant at Solihull, but the car was always dogged by controversy, a poor quality reputation, and finally by adverse currency movements between the British pound and the US dollar.

Production ended in October 1981, after 86,784 TR7 coupes, 24,864 TR7 dropheads, and a mere 2,815 TR8s had been built.

TVR

TVR

The origins of the TVR marque stretch back to the early 1950s, when Trevor Wilkinson set up the company in the famous seaside resort of Blackpool. The compression of his first name provided the initials TVR. Wilkinson ran the company until the begining of the 1960s, but a series of managers then saw the company struggling, and it was not until the Lilley family – father Arthur and son Martin – took over in 1965 that final stability was achieved.

The first series-production TVR was the Grantura, which became Mk III in 1962. This had a multi-tube frame, and a stubby fiberglass body shell, with a whole variety of proprietary engines available, and most were sold in kit form.

It was not until 1967 that the firm produced the first 'Lilley' TVR, the Vixen. Like the Grantura which it replaced, it had a multi-tube backbone chassis frame, and all-independent coil spring suspension, with rack and pinion steering. The engine was an 88hp Ford Cortina GT unit of 1599cc, which endowed the car with a 106mph top speed. Like its predecessors, it had a two-seater fixed-head coupe body made from fiberglass.

There were Vixens made until 1973, in four different types, all with the same Ford engine, and similar performance, though specifications, and detail styling, were improved gradually.

Following the vogue for fitting larger engines to sportscars, TVR first produced a Ford V8 powered Griffith in 1963 (with 4.7-liters and up to 271hp), which had phenomenal acceleration, though the quality was poor. Under Martin Lilley, this car's successor was the Tuscan V8 range of 1967-70, which in its long-wheelbase form was a much more driveable car, and better built into the bargain. With a recorded 0-100mph in 13.8 seconds, it was even quicker than the famous Cobra!

In 1969, however, TVR introduced the Tuscan V6, which was a more practical proposition for UK buyers, because it was fitted with the 2994cc Ford (UK) V6 engine, which produced 128hp. In most other respects, the Tuscan V6 was like the Vixen of the day, but had a top speed of 125mph.

After a further diversion into producing Tuscan-based cars with the Triumph Spitfire 1300 engine of 63hp, and the 2500 model, with a 106hp 2498cc Triumph six-cylinder engine (which was commercially more successful), TVR then turned to making a new 'M' Series of cars from 1972. Although still recognizably descended from the earlier TVRs, they had all-new multi-tube frames, and a sleeker body style. Cars were built with 1599cc Ford, 2498cc Triumph, and 2944cc Ford engines (1600M, 2500M and 3000M), the spare wheel was carried up front, ahead of the engine, and the trim was altogether more up-market and luxurious.

The most powerful of all was the Turbo, made available in 1976, this having a turbocharged V6 Ford unit producing 230hp, and giving the car a top speed of nearly 140mph. Only 63 Turbos were ever built. Also from the mid-1970s, came the Taimar, effectively a 3000M with a hatchback rear body conversion, and (in 1978 and 1979 only) the Convertible, which was an open two seater, also based on M-Series engineering, and the V6 engine.

In 1980, TVR launched a completely new range, called Tasmins. These were first offered as two-seater hatchbacks, but within a year a 2+2, and a convertible derivative, had also been launched. All cars used the German V6 Ford engine, which had 160hp with the help of fuel injection, and the top speed was approximately 130mph. There was sharp-nosed fiberglass bodywork, and a new design of multi-tube chassis.

This basic design has formed the basis of all more recent TVRs. Late in 1981 a 2-liter version of the car was announced (with a 1993cc Ford ohc unit), and a one-off 2.8-liter Turbo was shown in 1982, but after ownership changed again, a Rover-engined 350i (with 3528cc V8 engine) was launched, and this was later joined by the even more fearsome 390i, in which the Rover engine was enlarged to 3.9-liters, and 275hp.

Left: 1980 TVR Tasmin.
Right: 1983 TVR Tasmin 3.5-liter V8 leads the resurgence of the TVR company's fortunes as the sportscar itself gains ground.

Vauxhall

The marine engineering company, Vauxhall Iron Works, entered the car business in 1903, from London premises close to the Houses of Parliament. The business soon moved out to Luton, where the company is generally credited with producing Britain's first proper sportscar, the L H Pomeroy-designed Prince Henry model of 1913.

The name stems from the long-distance competitions, trials-cum-rallies, which were promoted in Germany at this time, for Prince Henry of Prussia, the Kaiser's younger brother, was an avid motorist, and gave his name to these events.

Pomeroy, who had joined Vauxhall in 1905, was a great designer, and always interested in sporting motoring. In 1910 he had persuaded a modified 20hp Vauxhall to reach 100mph at Brooklands. The Prince Henry used a similar engine, and was made in very small numbers before the outbreak of World War I. Its 4-liter engine had only four-cylinders, and side valves, but although performance was only modest, the general level of handling and control was superb by the standards of the day, and its looks set a yardstick by which all the next generation of sportscars were designed. It was easily recognized for having a tapered radiator, and the famed bonnet flutes.

At the outbreak of war, the Prince Henry had already evolved into the classic 4½-liter 30/98 model, and this was revived in 1919. As the side-valve model, the E-Type, it was built up until 1922, featuring such Edwardian niceties as exposed valve springs, and a fixed cylinder head. Only rear wheel brakes were provided, and it needed a brave driver to make use of its 60mph cruising speed.

Before L H Pomeroy left Vauxhall to work in the USA, he had been working on advanced engines, and in 1922 Harry Ricardo was involved in helping to produce new Tourist Trophy cars which had twin overhead camshaft engines. The later 30/98 road cars, the OE Models, however, had overhead valve engines, with pushrod operation, and this meant that peak power was boosted from 90hp to 112hp (ultimately 120hp). From 1923, front-wheel-brakes were available as an option, these being a rather strange design, with four shoes in the drums, which operated in conjunction with a transmission brake when the foot pedal was operated. The handbrake lever continued to operate the rear drums, independently of this. Various private-enterprise attempts were made to improve on this rather alarming system, which was not abandoned until hydraulic brakes were adopted on the last batch.

General Motors of Detroit bought the financially-ailing Vauxhall Motors in 1925, and soon began to transform it into a mass-production concern. Naturally the 30/98 did not fit into this strategy, and was phased out in 1927 after only 312 examples had been built in 14 years from announcement.

Despite their small numbers, these cars earned an enviable reputation for distance records (in Australia, a car completed the 565 mile Melbourne–Sydney run in 14hr 43min. in 1923), speed trials, hill climbs and club events. The 30/98, in any case, was a major influence on the layout of the first 3-liter Bentley, for one of its designers, Harry Varley, had been recruited from Vauxhall.

All 30/98s equipped with lightweight bodies were guaranteed to reach 100mph, and at least 85mph was possible, even on four-seater tourers. The aptly named Velox model weighed about 3100lb.

Since then, the Vauxhall company has never built a sportscar, though one ought to mention the two 'homologation specials' of the 1970s – the 'droop-snoot' Fierenza coupe of 1974-1975, and the 2.3-liter twin-cam engined Chevette HS four-seater hatchback, of which 400 were built between 1976 and 1979.

Below: 1914 Vauxhall Prince Henry.
Top right: 1923 Vauxhall 30/98.
Bottom right: 1927 Vauxhall 30/98 Wensum.

K 205

Index

Page numbers in italics refer to illustrations.